The Trials of an Expert Witness

By the same author:

Bone Idle on Parade – Memoirs of an incompetent soldier

The Trials of an Expert Witness

The indiscreet memoirs of a valuer and planner

Hilary Eve

Aeneas

First published in 2005 as a co-publication
by Aeneas Press and Estates Gazette Books

PO Box 200
Chichester
West Sussex
PO18 0YX
UK

Typeset in Plantin Light
by Marie Doherty

Printed and bound by
Antony Rowe Ltd
Chippenham
Wiltshire
United Kingdom

ISBN: 1-902115-46-5

British Library Cataloguing in Publication Data
A catalogue record of this book is available from the British Library

Eve, Hilary

To the memory of my father, Gerald

Acknowledgments

I would like to thank the following people:

- Ian Heritage for his research;
- Michael Hopper, John Polling and Richard Pratt for their help on the text;
- Peter Haddock, Michael Hopper, Nigel Rawlence, John Seifert and Donald Scannell for supplying photographs;
- Simon Eve for supplying copies of letters written by his grandfather, Gerald Eve.
- My wife, Mary, for her great help in reproducing all the illustrations, and for her infinite patience in helping me to use a computer.

Any errors or omissions are wholly my responsibility.

The Times obituaries of Professor AS Eve, Dr FC Eve and Lord Silsoe are reproduced by kind permission, and are the copyright, of News International Syndication, London 2004.

The obituaries of EC Strathon and JD Trustram Eve reproduced from Chartered Surveyor Weekly and the obituary of C Gerald Eve from the Journal of the RICS are reproduced by kind permission of the Royal Institution of Chartered Surveyors.

Contents

Appendices
Obituaries
Other

Illustrations

Preface

This book of memoirs is a personal account of certain incidents of my professional life, with occasional diversions. Much of it concerns my time at Gerald Eve & Co, and its publication fortuitously coincides with the 75th anniversary of my father's founding of the firm on 1st January 1930. Yet this book is most certainly not a history of the firm; nor is it in any way a balanced or (God forbid!) a complete account of what went on when I worked there. It is confined to what I can, or choose to, recall and is necessarily highly selective.

I have been critical of a number of people in this book, be they architect, planner, surveyor or lawyer. However, I would like to stress that in no way do I ever intend to impugn anyone's professional integrity.

I apologise to those of my former colleagues who are disappointed not to see their names in this book, but I readily acknowledge all the help they gave me – and thank them for the tolerance they showed when coping with someone who had an (not always expert) opinion on everything and was never afraid to express it.

Part I. *Gerald Eve & Co*

Mayfair, London

STUDENT

April 1948

Chapter one

No Name on the Door

It was 9.25 am precisely as I walked up to the entrance of 6 Queen Street Mayfair. I was making sure I was not going to be late for my first day at the office. I expected to see a brass plate with GERALD EVE & CO, CHARTERED SURVEYORS engraved upon it; I was not then aware that the founder, my father Gerald Eve, took the view that professional people should not advertise – not even on the front entrance of their place of work; so there was no brass plate – well, that's not quite true because on the door was a tiny one saying PUSH. Not unreasonably, I pushed, but it was a rather gingerly push because I wondered if I had come to the right place. Looking back on it today, some fifty-six years later, I think that probably I had not. However, that is something you will have to decide for yourself when you've come to the end of my story.

Author, 1948, aged 21

To my relief, there at the end of a rather dark entrance passage was a small illuminated sign which read: GERALD EVE & CO. RECEPTION. Shielded from the eyes of the general public, it was apparently permissible to have a notice inside the premises stating the name of the firm. I walked up to the reception kiosk and before I could utter a word an Eton-cropped woman, whom I was later to know as Miss James, said:

'Ah, good morning. You must be Mr Hilary. Mr Strathon is expecting you.' Thus was I received into the firm that my father started in 1930.

It was 23 April 1948 and my father had died just over two years previously. I had been demobilised from the Army only a fortnight and had arrived home in Weybridge from my unit in Edinburgh. There, as a Second Lieutenant, I had been in charge of a Signal Office entirely staffed by pretty ATS girls – and I had been going out with the prettiest one. The Army had provided me with: enough money to go out every evening with my girlfriend, a car, my uniform, a comfortable Officers Mess with all meals provided (waiters in attendance), a pleasant bedroom, and an Army batman who looked after my clothes, cleaned my shoes, tidied my room, brought me morning tea and ran my bath.

I had a desk job that was wholly undemanding and involved pushing paper into three different trays – IN, OUT and TOO DIFFICULT. The contents of the third tray were delivered to my Superior Officer for him to deal with. (I understand a past Secretary of the MCC also had three filing trays, but his were marked IN, OUT, and LBW – let the buggers wait.) When I gave orders they were obeyed. Wherever I went in Camp, I was saluted. All this before I was twenty-one.

When I got home I had to make my bed, tidy my room, clean my shoes and help in the house. No one saluted me – indeed my stepfather was a Captain RN so if there were any saluting to do, it would be me to him. The only orders I could give were to the cat which, as cats do, did the opposite. Now I had to commute to London six days a week (we worked Saturday mornings) and be the lowest form of life in the office. Things were going to change somewhat – and it didn't seem for the better.

Back at Reception, Miss James telephoned Eric Strathon, a partner whose office was in the front room of the ground floor. He came out and greeted me warmly.

'Hullo, Hilpy. I am glad you have come because we can do with a bit of help. We've masses of work on.' He was a lovely, kind man and his calling me Hilpy brought a lump to my throat as the only other person ever to call me that was my father.

EC Strathon, partner in Gerald Eve & Co and, later, Member of the Lands Tribunal 1969–80

'Mr Strappon', as I called him as a little boy, who joined my father at the start of the firm, had been made a partner by him and returned from the war as Major Strathon. He and his new bride, Peggy, had moved into our house before the war to act as parents to my four siblings and me while our parents were away abroad on holiday. We loved it and we loved them. When my parents had a dance in the house every year my two brothers and I used to wait with pillows ready for 'Mr Strappon' to come in to see us in our bedroom and have a pillow fight. He never disappointed us.

'You'll be working for Billy Oliver and Bill Haddock who are in the office next door. I'll introduce you,' he said.

We went next door and I met Billy and Bill. I found out later that Billy's real name was Walter;

WF (Billy) Oliver

understandably, he didn't want to be called a 'Wally'. Bill's real name was Percy; also understandably, he certainly didn't want to be called that. If only we could choose our own names – I used to get beaten up regularly at prep school for having 'a girl's name'. Thank you mother, dear.

PGE (Bill) Haddock

Billy and Bill were seated on opposite sides of a large, oak partners' desk stacked high with fat, bulging files tied up in pink tape, with a little clear space in front of each chair. The two were nearly the same age as 'ECS', as Eric Strathon was called in the firm, and had joined him very soon after the firm's inception. I was given a seat at a small table set against the side of the desk, and Billy, unable, I suspect, to think of anything he could usefully give me at that moment, pushed over to me an incredibly fat and tattered file and said:

'Please read through that and let me know the present position.'

The file was entitled *Church House, Westminster, Claims under 2(i)(a) and 2(i)(b) of the Compensation (Defence) Act 1939*. At the foot of the front cover was written: *Instructed by Dr Geoffrey Fisher, Bishop of London.*

That rang a bell with me because I recalled that every time my father met Dr Geoffrey Fisher he would come back with several very funny, risqué stories which the Bishop had told him. (Those stories probably helped him later to become Archbishop of Canterbury.)

I started flipping through the file quickly but, hard as I looked, none of the Bishop's risqué stories was recorded on the file. Today, it seems to me that those involved had little sense of what it was important to record and what was not. After all, Gerald Eve & Co were meant to be valuers, and jolly good ones at that, but they seemed totally unaware of what would be valuable to record for posterity. Thus, due to their mistaken sense of values, I am unable to retail to you any of Dr Fisher's very funny, risqué stories. Which is a pity. The file was simply full of boring schedules of dilapidations and even more boring lists of fixtures and fittings which had been lost or damaged while Church property had been requisitioned.

My boredom was soon to be relieved by a knock on the door. In came a rather spivvy-looking man with Brylcreemed hair and a thin moustache that looked as if it had been pencilled on. I learnt later that it was George Offord, a brilliant cockney plant and machinery valuer who had left the firm, probably because he drank too much.

'Ullo me ol' cock sparrers, 'Ow's, tricks?' said George. He looked behind Bill Haddock and caught sight of a photograph on the wall with a caption: *Charles Gerald Eve. President of the Chartered Surveyor's Institution, 1932–1933*. He said: 'I see you've still got that old bastard 'anging up on the wall.'

'You'd better be careful what you say about him, because this is his son, Hilary, sitting right here,' said Bill quickly.

'Oh really,' said George, not batting an eyelid, 'Pleased to meet you 'Ilary. I always 'ad the greatest respect for your father.'

CHARLES GERALD EVE, Esq.

PRESIDENT OF THE CHARTERED SURVEYORS' INSTITUTION, 1932-1933

Charles Gerald Eve, Esq., President of the Chartered Surveyors' Institution, 1932–1933

Chapter two

The Belles of Stepney

Billy Oliver told me he was going to a meeting of the Assessment Committee at Stepney Town Hall after lunch and said I could accompany him.

'I think you might find it a bit of an eye-opener,' he said.

When we got onto the platform of Green Park Tube Station, Billy walked a certain distance, stopped abruptly and turned towards the railway line.

'This is where we need to stand on the platform to be opposite the exit at Victoria where we get out to change onto the District Line,' said Billy. He proved to be exactly right. The same procedure happened at Victoria and, sure enough, the carriage doors opened precisely opposite the exit sign at Whitechapel, our destination. I was impressed.

'Mr Oliver, do you know where to stand on every platform to alight opposite the exit at every Tube station?'

'I wouldn't put my shirt on it; I might fail on some of the more outlying stations, such as Stanmore; though I do actually know Stanmore, because it's at the end of the line, so you have to get on at the very front so that you can pass through the exit gate next to the buffers.'

After quite a walk, we entered Stepney Town Hall and went to a waiting room where the Clerk of the Assessment Committee met us. He said he would go and get his committee assembled to hear our case.

In those days it was the local Council who assessed all the rateable values for every property that was charged rates. If you thought your assessment was too high you appealed to the Assessment Committee of the Council concerned. If you didn't like their decision you appealed to Quarter Sessions which, I suppose, today is probably the equivalent of a Crown Court. A judge heard your case; there was no jury.

The Clerk returned and said the Committee was ready to hear our appeal. He led the way into a vast room where at the far end were about a dozen people seated behind a horseshoe-shaped table. In the middle of the room, like a little desert island in a vast ocean, was a small wooden chair similar to what I used to sit on at prep school. Billy put his briefcase on the floor beside the chair.

'We wasn't expecting you to bring anyone else, Mr Oliver, but the Clerk will find another chair for the young man, won't you, Fred?' said the man, wearing a bright red tie, seated in the middle of the horseshoe. Fred went out and came back with another chair equally small, which he placed close up to the other one. We both sat down, looking like two Babes in the Wood.

Billy fished out some papers, gave some for me to hold – there was no table to put anything on – stood up and opened his case. He explained, very clearly and convincingly I thought, why the assessment of this factory was excessive.

I noticed during the proceedings that at the far end of the horseshoe were two hatted women who never stopped talking to each other. Each had in front of her a piece of knitting which lengthened visibly during the presentation of Billy's case.

After a final impassioned plea for a reduction in the assessment, Billy sat down and the Council's Rating and Valuation Officer was asked to reply to Billy's case. To be fair, I must record that the two hatted, talkative, knitting women treated Billy and their own valuer with complete impartiality – at least at this stage – as they did not cease to talk during their valuer's representations.

The Council's valuer sat down; the Chairman looked along either side of the horseshoe raising his eyebrows; successive members shook their heads; it was unanimous except for the two knitters who were so absorbed in their conversation that they did not notice their Chairman's enquiring glance.

Ten votes against any reduction in the assessment and two abstentions, recorded the Committee Clerk. He passed a note to the Chairman who looked down at it and cleared his throat. He read out its contents haltingly, stumbling over some of the longer words:

'After hearing the representations of both sides, and after giving the matter due consideration, the decision of this committee is that the assessment be confirmed. Thank you, Mr Oliver. I now declare the meeting closed.'

Billy stuffed his papers back into his briefcase.

'Let's f*** off out of here, he said, none too quietly. 'Perhaps we can get some justice at Quarter Sessions.'

We went to a local café and had a cup of tea, the strongest I had ever tasted, and a stale curled up sandwich spread with revolting margarine.

'The trouble here, Hilary', said Billy, is that they are a bunch of Commies and don't give assessment reductions to capitalists. Their local MP, for Mile End, is Phil Piratin, who is a Communist; so are some of the Councillors.'

It was no surprise to me a few months later when the Government passed the Local Government Act 1948; this took the task of making rating valuations out of the hands of the local authorities and into the hands of the Inland Revenue. The Council's job was confined to collecting the rates.

Appeals were to independent Local Valuation Courts, which did not have local Councillors as members. Appeals from the Court's decisions were to the Lands Tribunal whose members were barristers or chartered surveyors. Later on in my career I was to see much of these two bodies.

Chapter three

Brainwashed!

Before I go any further, I had better tell you why I joined Gerald Eve & Co in the first place. It was because I was brainwashed into it. That's a fact. From my earliest days (I was born in 1927) I had been told that I was going to go into Daddy's firm to be a surveyor, just like Daddy. This seemed OK to me. In the early thirties, Daddy went off every morning in a chauffeur-driven car to Weybridge Station and was usually back in time for dinner. Once I remember my mother getting into a bit of a panic because Gerald rang up to say that, unexpectedly, the Countess of Ayr was coming to dinner. My mother telephoned the butcher, the greengrocer, the fishmonger, the grocer and the wine merchant and soon a succession of errand boys were bicycling up the hill to our house in Weybridge Park. The hill was so steep that the errand boys would stop their traditional whistling and, swearing under their breath, have to get off their bikes and push.

Cook soon worked her magic on the newly delivered provisions and, by the time the front door bell rang, a dinner fit for a princess, let alone a mere countess, had been prepared. My mother looked in the mirror in the hall to tidy her hair – it was already as tidy as it could be – and put on her very best smile as the parlourmaid opened the front door. My father stood next to a scruffy little man in a dirty raincoat.

'Darling, this is Mr Jones, the County Surveyor.'

I suppose one reason I went along with being a surveyor, rather than a Prime Minister, a film director, an advertising executive, a barrister or a psychiatrist, which at times were my other inclinations, was that it seemed to provide a pretty comfortable existence. We had a swimming pool built in about 1934 and nobody else in Weybridge Park had one, so I felt we were pretty lucky. In fact we weren't that lucky because the water in the pool was ghastly at times. Once I remember swimming in the pool when it was dark green and had hundreds of little frogs in it. Occasionally, huge glass carboys of chlorine, protected by a straw lining inside a metal framework, were delivered to us and poured into the pool. Nonetheless, I spent most of my time during those three years that we had the pool suffering from continual sties (38 of them) in my eyes and abscesses in my ears.

When I tell you about the Eve family you will realise why I was destined to become a surveyor. The first Eve I will tell you about is not the one in the Garden of Eden but Mr Justice Eve. I am not sure what relation he was, very distant I think, but I am told he was very fat. He had had a good lunch and was hearing a very long and boring case about a disputed will; counsel was reading out a long list of blood relatives. Eve lost his patience:

'Yes, Mr Smith, we all know about blood relatives. Some are bloodier than others.'

The Eve family was no exception. I get bored when people go on about their ancestors so I will only tell you about those that were larger than life – which was most of them.

One reason why the Eve family was so well-known among chartered surveyors was that unlike other surveying families, like, say, the Cluttons, the Eves found by bitter experience that it was better to have only one Eve per firm, unless they were father and son. So there was Richard Eve, who was in practice doing rating valuations in Silsoe, Beds, in 1812 – a time when Napoleon Bonaparte was having a bit of trouble around Moscow; Richard Eve & Son (founded 1856); William Eve & Sons (1860); Page, Harding & Eve (1865); JR Eve & Son (1893); Thurgood, Martin & Eve (1920); Gerald Eve & Co

(1930); and Brian Eve (1943).

When JR Eve & Son celebrated in 1965 the centenary of the birth of Herbert Trustram Eve who was a partner in the firm for forty-three years (1893-1936), they produced a brief history of the firm. This included a family tree, complete with serpent and apples, of Eve Surveyors, who originate from the Rodings of Essex.

Richard Eve of Silsoe, my great-grandfather, may have been a chartered surveyor because,

THE HISTORY OF EVE SURVEYORS.

The History of Eve Surveyors reproduced from *The History of J R Eve & Son* (1965)

although the Institute of Surveyors (as the RICS was then called) was not founded until 1868, when he was aged eighty, he never retired and probably never ceased to practise until his death in 1885. (It was John Clutton who took the lead in gathering together forty-nine leading surveyors in 1868 to form the Institute.) Richard was born at Baldock in 1788 and at the age of eighteen moved to Silsoe, Bedfordshire where he rented West End Farm from Baroness Lucas. He was the tenant there for seventy-nine years, from 1806 until he died aged 97, having had two wives and seven children. At the time of his death he had a hundred and one descendants. His obituary in the local paper said he 'attained wide celebrity as "The Senior Churchwarden of England" as he was often called, and probably with justice, for it is hardly likely there is another who has served that office for seventy years'.

In the obituary of Pearl Pleydell-Bouverie in 1996, *The Times* newspaper suggested that she held the record for Churchwarden by serving for 65 years; but the very same day the then current Churchwarden of St James Church, Silsoe, wrote to inform readers of, and to defend, his predecessor's record. His defence appears to be successful, for no one has written to claim a longer period of service than Richard Eve. It used to be in the Guinness Book of Records.

Chartered or unchartered surveyor, Richard Eve was certainly a farmer, and also a valuer for, according to his obituary, he 'resigned his membership of the Board of Guardians to undertake the valuation of Ampthill Union'. A 'Union' comprised a number of parishes combined under one Board of Guardians for poor law administration. The job entailed, I presume, putting a rating assessment on every rateable property, which in those days included dwellings, in Ampthill Union.

Both Richard's sons were chartered surveyors, One, John Richard, stayed behind to look after the firm of his old dad who died in 1885, while the other, William, went to London in 1851to be trained in the profession. In 1860, he set up a firm, William Eve & Son, and made a fortune – £250,000 I was told – practising as a surveyor; that was quite a sum in those days because William died in 1916, aged eighty, and he probably retired some years before then. One of the reasons he had so much money may be because he did not spend any; for the tale my mother told was that no one in the family was allowed to light the candles in the drawing room until father William raised his hand. His wife and daughters had to continue doing their sewing, or whatever, until it was nearly dark.

John Richard Eve
(1833–1902)

John Richard bought into a partnership in 1865 and eventually set up his own firm.

He had two sons who were surveyors: a partner and Gerald.
Herbert, who was seven years
older than Gerald, was taken
into partnership in 1893.
Gerald, the youngest of seven,
was articled to JR Eve & Son in
1891, but Herbert told him he
could not be a partner in the
firm. According to my mother,
Gerald was 'robbed of his
birthright.'

Sir Herbert Trustram Eve KBE
(1864–1936)

Gerald qualified as a chartered surveyor in 1894 and it
seems that he spent some time with William Eve & Son, for I
have a copy of a letter he wrote in 1895 to his father on William
Eve & Son writing paper, describing a Quarter Sessions hearing of a rating appeal which he and William attended.

Gerald's diary of 1896 indicates he worked for Messrs Body
and Son at Plymouth and it seems he was a pupil there.

In 1899, Gerald got a job as Land Agent to the Kitley Estate
near Yealmpton in Devon. In June the year before – this
transpired to me quite late in my life – Gerald at the age of 24
had at Okehampton, amazingly, married a widow with five
children.

The name of the family that owned the Kitley Estate was
Bastard. Some people, meeting the Bastards for the first time
and embarrassed by the name, pronounced it by accenting the
second syllable of the word – Bas**tard**. No said the Bastards, the
name is **Bas**tard. They were very proud of their name because
they were descendants from a bastard son of William The
Conqueror.

I have a copy of a letter Gerald wrote to his father in June
1899 telling him about the estate and asking his advice about
the use of corrugated iron. It makes you realise how different

2 July 1895

From
WILLIAM EVE, F.S.I. & M.N.
Surveyors, &c.,
41, UNION COURT.
OLD BROAD STREET, E.C

Dear Pater,
The Assessment Committee was
all round re Newport Pagnell.
The Chairman of Quarter Sessions began
by asking our Counsel "if we can settle
the Gross, can we from that, with out
argument, Deduce the Ratable." Our
Counsel said 'Oh yes— there is a customary

Typescript of letter, dated 2 July 1895 from Gerald Eve, aged 21, to his father:

Dear Pater,

The Assessment Committee was all round re Newport Pagnell...

Mr Eve [William Eve, Gerald's uncle] says Roger made a fool of himself, made a long speech as if he were in the Rostrum. Counsel cross examined him (prompted by William Eve) – his evidence seemed to have been like Cumberlands over Green's Brewery – he had been over the place and formed an opinion – but had no details and knew nothing about plant. Had cubed the \buildings up but had done nothing with the result. Reader was their best witness – very shrewd, but knew nothing about rating.

At the close of the Appellant's case Mr Eve suggested to counsel to say "there was no case to answer and did the Court think it necessary to call any witnesses for the Committee." Court said "no"... except one man, so they called Arnold (?). who had made the Valuation –

Thus the cases were won on cross examining entirely.

Mr Eve's account is this:

Hipwell's Brewery re				
Brewery Valuation and Report		12	12	0
5 Maltings and Report		2	12	6
Attending at Aylesbury	)			
Engaged 12 hours	)	4	4	0
Expenses		2	10	9
		21	19	3
Allfrey & Lovells Brewery				
Brewery Valuation and Report		8	8	0
2 Maltings do.		1	1	0
Sending for Ordnance Sheet and enlarging a plan ? [illegible]		1	1	0
Exs.			4	0
		10	14	0

(This case was not tried)

I forgot to tell you that I saw Ellis in Bermondsey, who used to be a pupil at Silsoe; he is partner in a large Tannery. Said he recognised me from my mother! Sent his kind remembrances to all.

Your affect. son, Gerald

The Manor Office.
Yealmpton.

6 June 1899

Letter dated 6 June 1899 from Gerald Eve, at The Manor Office, Yealmpton, Devon, to his father:

'Dear Pater,
I cannot remember whether it is 22 or 24 gauge you recommend for corrugate iron. Do you generally close board underneath C. Iron roofs. If not do you get many complaints of leakage? It seems to me that Iron plus close boarding is as expensive as slating.

The rental here is between £9000 and £10,000, and there are other receipts. Altogether there are 300 tenantry (no allotments). There is a regular accumulation of work (chiefly schemes) left for me. The weather is fearfully hot and...'

things were in those days when telephones were not yet in common use. It meant the pace of life was so much slower than today.

Gerald was an early motoring enthusiast for I remember him telling me he bought a Renault motor car in 1906.

Another thing he told me about those days concerned some pet white rabbits which roamed at large round the Bastards' house. Pearl, the gamekeeper, walked up the long drive to the house and told Gerald that some of the pet white rabbits were getting caught in the snares he was setting.

Gerald told the story as told by Pearl in a Devon accent:

'So Mr Eve said: "Don't worry, Pearl. I'll solve the problem. Come back to me this afternoon and I'll 'ave the answer. So I goes back to him after dinner and 'e gives me four small white notices attached to little stakes, and said: " Stick one of these in the ground beside each snare". Each notice had written on it: THESE SNARES FOR GREY RABBITS ONLY, but I 'adn't walked to the end of the drive with 'em before I realised 'e was 'aving me on.'

That is a true story; I am not so sure about this one, which again has to be told in a broad Devon accent:

George was walkin' along the road one day when 'e saw 'Arry standin' in the field.

''Arry', 'e says, 'are you comin' fishin' with me today?'

'No', says 'Arry, so George goes fishin' by 'isself, see.

The next day George is walkin' along the road and 'e sees 'Arry still standin' in the same place in the field and 'e says to 'im:

''Arry, are you coming ferretin' with me today?' and 'Arry says:

'No', so George goes ferretin' by 'isself', see.

The third day, George is walkin' along the road and he sees 'Arry again standin' in the same place in the field. Says George:

''Arry, are you comin' poachin' with me?'

'No', says 'Arry. George says:

''Arry, why don't you go any place with me?' And 'Arry says:

'I've got my foot caught in a rabbit 'ole.'

Around 1906 or perhaps earlier, Gerald and his wife split up and he told his brother, Herbert, that if he took her into his home he, Gerald, would never speak to him again. Herbert took her in (apparently she was a friend of his wife) and Gerald indeed never did speak to him again. This was somewhat inconvenient if not embarrassing at times, for when they were negotiating across the table, Gerald would speak to Herbert's junior partner, Mark Wilks, and Herbert would do likewise.

Sometimes they were expert witnesses in Court on different sides and one can imagine the rivalry between the two. On one case, concerning the rating assessment of Bentalls, the Kingston-on-Thames department store, the judge was Mr Justice Eve; the expert witnesses were Herbert and Gerald; and Gerald's counsel was Malcolm, Herbert's son. I do not know if Bentalls won their case, but certainly the Eve Family couldn't lose, as they had backed it three ways.

Herbert's second Christian name was Trustram which was their mother's maiden name and he decided to add this to his surname. For some reason people always remember the name Trustram Eve – perhaps it indicates they are to be trusted – and Gerald said the name was worth £10,000 a year – quite a sum of money in the twenties and thirties. Herbert was knighted (K.B.E.) in 1918 for his war service as Chairman of the Forage Committee of the War Office, which included the control of the hay crop requisitioned for the Forces in World War I. However, my father told a very different story: he said Herbert bought the honour from the Prime Minister, Lloyd George, who reportedly was not above selling titles in return for political donations. Perhaps we should take such an accusation with a pinch of salt in view of the strained relationship between the two brothers.

I know exactly when Gerald left his job at the Kitley Estate because I have in my possession a silver cigarette case that has engraved on the front: PRESENTED TO C. GERALD EVE ESQ BY THE EMPLOYEES on the KITLEY ESTATE, OCTOBER 1909.

Gerald left to join the Inland Revenue's new Valuation Office as one of thirteen Superintending Valuers (SVs) whose job it was to value the whole of the country in preparation for a new Land Tax contained in Lloyd George's strangely titled Finance Act (1909–10) 1910. He was SV for the Eastern Division and was the youngest of them all. I have a picture kindly donated to me a while ago by Mr Stedman when he was District Valuer, Guildford,

Chairman, Deputy Chairman and Principal Valuers of the Inland Revenue Department. Somerset House, June 1912

Gerald Eve, from the photo above, (standing 3rd left)

of them assembled outside Somerset House with their superiors in June 1912.

A close scrutiny of the photograph reveals that my father (back row, third from left) has no visible means of support. Where are his legs? The answer, I deduced, is that the person who sat in front of him was absent at the time; the photograph of him (seated on a different, upholstered, sort of chair) was put in afterwards with suitable touching up of the surrounds – apart from Gerald's pair of legs.

Gerald was based in Nottingham and he was in charge of the valuations of Leicestershire, Lincolnshire, Norfolk, Northamptonshire, Nottinghamshire and Rutland. In Nottingham he met my mother and they moved down to Surbiton in 1919. In 1920 Gerald went into a ten-year partnership (partnerships for fixed terms were common in those days) with Thurgood and Martin, a firm of chartered surveyors in Chancery Lane. Not surprisingly, the name of the firm was altered to Thurgood, Martin and Eve. After the ten years was up, Gerald decided to set up on his own account – he was getting in most of the work but his share was only one-third of the profits. Thus was born in 1930 Gerald Eve & Co of Chancery Lane.

Shortly after I started work at Gerald Eve &Co, Eric Strathon started making arrangements for me to be articled out. Gerald had told Eric that he would like me to start as a land agent, as he did, and had named two firms of land agents – Bidwells of Cambridge, and Rawlence and Squarey who had offices at Salisbury, Southampton and Sherborne, with a small office off Victoria St in London. Eric chose the London office as it was convenient for me to commute there from Weybridge.

The Articles of Apprenticeship dated 8 November 1948 witnessed *that the Clerk* [that's me] *of his own free will* [oh yes? I was brainwashed] *by these presents* [what presents? I never got any presents] *does place himself Pupil and Clerk to the Firm* for two years whereby I was to pay £150 for the first year and £100 for the second. I stayed on for a third year for nothing, at the end of which they were kind enough to give me a £300 bonus. I like to think I must have been useful by then.

I was fortunate in being able to obtain the money for my articles – my father had left his estate equally between his five children, but sensibly the will provided we were not allowed to get our hands on the money till we were aged 28, except by permission of the trustees who were the Public Trustee and my older sister, Diana. There was also an educational fund which would not be distributed to the five of us until we had all been educated, and it was this fund that stumped up the money for my articles.

Part II. *Rawlence and Squarey*

Victoria, London

ARTICLED PUPIL

8th November 1948

Chapter four

A Time Capsule

The Senior Partner of Rawlence and Squarey's London Office was Duncan Rawlence who seemed to me to be very old. He had probably come out of retirement during the war to keep the firm going and had been joined recently by his nephew, Michael Rawlence, to whom I was to be articled. Known as Mike, he had been demobilised from the Irish Guards, got a BA at Cambridge in Land Economy and was a recently qualified chartered surveyor.

Michael Rawlence

Uncle Duncan was to my mind very old-fashioned. For instance at lunch-time, he would sit at his desk and produce an old scratched Mackintosh's Quality Street Toffees tin with pictures of crinolined ladies on the lid. Out of this he would produce a hunk of bread and a hunk of cheddar cheese. He would then get out his penknife, open out the large blade and cut off pieces of cheese. These he would spear with his knife and take to his mouth. I soon learnt that he was not to be disturbed during this rather Spartan meal.

His other peculiarity was that when we went round farms together, my taking notes from his dictation, he would say 'one dialect cowshed'. I wrote this down religiously, not wishing to air my ignorance in not knowing what a dialect cowshed was; when I got back to the office, I asked Mike what it was.

'Oh,' said Mike, 'you had better get used to the fact that Uncle Duncan's "dialect barns" are derelict ones.'

I soon learnt that Duncan was not one of nature's big spenders and Mike had an awful job trying to get him to spend anything on modernising the office. The whole office was a time capsule not just of pre-war days, but it seemed to me to smack more of the twenties.

Let me put you in the picture. The general office had brown linoleum on the floor, once-cream walls which had been last decorated sometime before the war (probably the Crimean), bare bulbs with frosted glass shades above, and no curtains. The furniture comprised two old oak desks, two old oak plan chests and a drawing board. All along one wall were black tin boxes, stacked up to the ceiling, with names of landed gentry painted on them in white, such as 'Lord Willoughby de Broke'. Lord knows how long it took to get out one of the boxes at the bottom, but then I suppose, by the natural order of things, the most used boxes would make their way to the top.

The general office staff were a curious lot. There was nice Mr Yeo, who was a building surveyor – and a very knowledgeable one too. The trouble was that he was almost stone deaf but he put this fact to good use when he was asked awkward questions in the witness box.

'Sorry, I didn't catch that,' he would say, cupping his hand to his ear, and by the time the question had been repeated slowly and loudly, he would have had enough time to come up with a good answer.

Sometimes he would send me out to Stanford's in Covent Garden to buy an Ordnance Survey map. I would be asked to colour pink on the plan the extent of the farm we were dealing with. It was nerve-racking doing the colour-wash as, if you coloured in the wrong fields, you had to go off and buy another map. Yeo taught me how to make a really even coloured wash:

'Put the map on a slant; start the wash at the top; keep the bottom bit wet all the time; once you have started a colour-wash, don't stop, even if the Queen comes in,' Yeo would say. Fortunately, she never did.

Then there was Mr Fortescue, a charming, kindly man who had only one leg. I cannot recall how he lost it. He came in on crutches every day and presumably did not have enough of his leg left to be able to have an artificial limb fitted. His job, believe it or not, was to make a summary in the letter-book of every letter that went out from the firm. He did this in beautiful flowery handwriting.

Finally there was Mr Cooper, the accounts clerk, a mean man who would put himself out for nobody. When I offered him a cigarette he would say 'Ta very much', put it behind his ear and carry on working. He never gave me a cigarette. His aim in life, I am sure, was to be the most respectable man on earth. He would purse his lips and 'tut, tut' at practically everything, just to ensure he preserved his respectability. He also thought himself a great philosopher and at frequent intervals would come out with the tritest of trite expressions:

> It never rains but it pours.
> These things are sent to try us.
> These things come in threes.
> Worse things happen at sea.
> Oh, well, it could be raining.

And when he wanted to be really risqué:

> If you can't be good be careful;
> If you can't be careful, name it after me.

Cooper's job was to keep the firm's accounts and those of all the estates we looked after. All entries were handwritten.

In the passage was what I think may be described as a letter press. I don't know how it worked except that crinkly paper and purple ink were put into it and then the big screw at the top was turned to apply pressure to whatever was inside. In this way, copies of a letter were produced by Duncan's aged secretary.

Mike soon took on a secretary for himself and introduced carbon paper to the firm. This was adopted by Duncan's secretary, and slowly the firm started to be dragged into the twentieth century, although by this time we were almost halfway through it.

Mike managed the Wythes Estate at Epping, Essex, and, in return, was given the use of a delightful Georgian house on the estate. Mr Wythes at one time had a house in Grosvenor Square. The reason he moved was that he caught sight of a red bus going down Bond Street and so thought it was time to move on.

Mike was an extremely competent land agent and we had no end of fun going round the estates we managed. Sometimes we had to put up the rents – they varied in those days between £1 and £3 an acre – which sometimes made the farmer rather angry, though Mike was never unfortunate enough to undergo my father's experience. Apparently Gerald had, after some argument, agreed a much higher rent with a farmer who suggested they might seal the bargain with a glass of home-made cider. Gerald accepted and the next thing he remembered was waking up in a ditch in the dark.

There is a tale of a surveyor who was fond of his drink. Apparently, he was making an inventory of a house and when he looked inside an inglenook fireplace he found an old bottle of something alcoholic. He was found sometime later asleep on the floor and the last entry in the inventory was: *one revolving dining table*.

Rawlence and Squarey was a very old-established firm of land and estate agents. Indeed, in 1951 Mike and I went down to their head office at Salisbury, Wiltshire, to attend the dinner given to celebrate the firm's hundredth anniversary.

On his return to London, Mike decided to try to boost the estate agency side. He gave me the name of an applicant who wanted some offices. We did not have any on our books so I

Rawlence & Squarey's Centenary Dinner, 1st February 1952.
The author is in the front row, second from right.

rang round to other agents who sent us particulars on the basis of sharing commission. We had no headed paper for printing particulars on – and no means of copying them, so I cut all the heads off the particulars and then sent him the ones I thought might suit him. No dice.

A while later, a firm of agents said they had an applicant for some offices and asked if we had anything suitable. I sent them all the particulars I had left. I had a bit of bad luck then, because the agents wrote back somewhat indignantly to say that I had just sent them the particulars they had sent me a month earlier with their name chopped off the top. A bit embarrassing that.

After that, Mike decided that we had better take on someone who was an experienced estate agent. Along came Mr Gifford to take over the estate agency side of the business – if you could call it a side; it was more a rock bottom. The man was about sixty and had a very red face. He smelt of alcohol – which as it turned out was not surprising. Whether or not Mike took up any references concerning him is a mystery, but if he did the chap must have written them himself. The first thing Gifford did was to tell Mike he was buying a radio on a hire-purchase agreement and would he be kind enough to sign the form as guarantor. Mike duly obliged.

After a short while, the head of our estate agency department – indeed the entirety of it – after keeping erratic hours and turning up obviously drunk, failed to turn up for work. Mike and I went to his flat nearby and found that he had disappeared from there also, owing rent. To cap matters, a couple of months later, Mike had a letter from the hire-purchase company to say that Mr Gifford had failed to make any further payments on his radio after the initial one and that Mike must therefore pay the remaining ninety per cent.

We came back to the office and Mike told Cooper to make the necessary payments under the hire-purchase agreement and explained why. That was bad enough, but I really felt sorry for Mike when Cooper said all too predictably: 'Oh well, these things are sent to try us.'

As soon as we went out of the door, Mike said to me:

'Sometimes, Hilary, I wonder what I have done wrong for the Almighty to decide to send Cooper to try me. It must have been something pretty awful.'

We used to do tenant-right valuations for farmers, and Mike would plunge his arm into a haystack, pull out a bunch of hay, put it to his nose and inhale deeply. 'First Quality' he would say to me and I would record it dutifully in my notebook.

When we walked the fields with a farmer, every time Mike asked him a question, the farmer would stop to answer him. It seemed to me that farmers had not mastered the art of talking and walking at the same time.

One day, Mike announced: 'Tomorrow will be an XK120 operation.'

He had just bought a Jaguar and was dying to go for a decent spin in it. It was a wonderful trip and extremely exhilarating when Mike came across a bit of open road and put his foot down on a motor obviously rarin' to go. We arrived in Gloucestershire in a remarkably short space of time. As we drove up Birdlip Hill, which is pretty steep, we came up behind a Reliant Robin, a three-wheeler, which was having some difficulty getting up the hill. As we closed with it I was just able to read the notice on the back windscreen: 'Life is an uphill struggle without Jesus'.

We had to value a farm attached to a lunatic asylum, as it was called in those days. On the way down, Mike told me his lunatic asylum story.

A landlord took over a pub next to an asylum. The first day a man came in and said he was the director of the asylum. Would the landlord mind if some of his harmless inmates came and used his pub? The only thing was that he did not trust them with money so they would pay for their drinks with beer bottle tops. He would then settle up at the end of each month.

The landlord said that would be fine, and during the next month he did a roaring trade with the inmates from the asylum. At the end of the month, the chap from the asylum came to settle up. The landlord said:

'I can't tell you how grateful I am for your putting all this business my way.' He produced a sack. 'I have here 848 beer bottle tops which means you owe me £212.'

'Fine', said the chap from the asylum. 'Have you got change for a dustbin lid?'

When we arrived at the asylum, the manager told us that some of the inmates would be working in the fields but that we would be all right so long as we humoured them. He advised us not to turn our backs on them.

Off we went, walking the fields, and soon a man stopped his hoeing, came up to us and said very pleasantly.

'Good morning. I am Lord Blanchester. I own all this land round here as far as the eye can see.'

Mike, with the manager's cautionary advice in mind and keenly aware of the hoe the man was holding, said: 'Oh really? You're a very lucky man.'

The man turned on him angrily, raising his hoe and shouting: 'Oh no I'm not! It's nothing but a bloody nuisance.'

Mike quickly issued a stream of apologetic, soothing and sympathetic pleasantries as we both started walking slowly backwards. You may be sure that for the rest of our stroll round the farm we gave any farm-workers we saw a very wide berth.

Mike's real interest was in the City and what he called 'big deals'. When he went to a cocktail party, instead of chatting up the pretty girls, as I would, he would talk to chaps whom he thought might be good business contacts, and make notes in a tiny notebook he kept on him. Often he'd come into the office on a Monday morning with a brilliant idea.

'Do you know, Hilary, I met someone at a cocktail party on Saturday night who grew tomatoes in Devon. He found he was paying a king's ransom buying bamboo stakes for them so decided to grow bamboos himself. Now, he's made an absolute fortune in bamboos and has given up growing tomatoes. Find out what sort of climate bamboos like. Do they need a lot of rain like they get in the West Country, and do they need a mild climate as they have in Devon because of the Gulf Stream?'

Not long after, Mike rushed in with another possible 'big deal':

'Hilary, I have been offered a bit of land in Rio de Janeiro harbour which sounds really cheap. At the moment it's just a bit of mud most of the time, but my chap has heard very privately through the grapevine that the harbour is soon going to be dredged. This will enable big ships to come into this part of the harbour and my bit of mud could be worth a fortune as a wharf. The price is two million cruzeiros. Would you please find out, as a matter of urgency, what the Brazilian exchange rate is?'

Mike, inevitably, ended up in the City and made good. He told me he had gone into a consortium to build an office tower block on the south side of the Thames next to Putney Bridge and had made enough money to educate both his children, his son being put down for Eton.

He came into the office one Monday morning and said: 'Bit embarrassing on Saturday night, Hilary. I was playing poker with the local GP and we bid up rather a lot on one hand, neither of us being willing to stop. We both kept saying to the other that he really shouldn't raise the stakes any more, but neither would give way. When we got to the point where the doctor hadn't any more money to stake he said he'd throw in his Jaguar car, which he did. I thought I'd better stop bidding at that stage as the doctor, short of selling himself into slavery for the rest of his life, had nothing else to stake. I said I'd see him, so he plonked down three jacks .I capped them with three queens. That meant he had no car to get round to his patients today.'

'Crumbs! What's he going to do today about his morning round?' I said.

'Yeah, that's what I thought, so I played poker with him again yesterday and made sure I lost his Jaguar back to him – but not before I had had a wizard drive in it yesterday morning. That car certainly can move.'

One of our clients was Berry Bros and Rudd, the Queen's wine merchants, whose shop is at 3 St James's Street. When Mr Rudd died we had to value the firm's properties for death duties.

The shop was the most intriguing I have ever entered. It was a Georgian coffee shop and the coffee was weighed on some large scales. These scales had a seat fitted to them and customers used to come in and weigh themselves. Most of the interior has been preserved as it was then and the scales and book containing their customers' weights are still there today. You can see how much weight characters such as the Prince Regent and Beau Brummell put on during the years. Apparently, where

Berry Bros. and Rudd really came into their own was when there was a fire at the King's Palace in Whitehall and the King and his Court moved into St James's Palace, which is but a few yards away from their shop.

Their other property comprised some railway arches in Southwark. Mike and I went there to value them; at the end of our work, about 2 pm, the warehouseman offered us a glass of claret. The glasses he gave us had stems but no stands on the end, so, inevitably, as we couldn't put the glasses down any-where we drank the bottle rather quickly, thanked the man and returned to the car. Mike drove his car, a pre-war Studebaker, somewhat erratically back to the office. Once in Mike's office we felt very tipsy and got the giggles. We were quite incapable of doing any office work and, I fear, descended to rather a child-ish pastime. We looked up in the telephone directory people whose name was Smelly or Smellie and rang them up. When they answered we said:

'Are you Smelly?'

'Yes', they said:

'Well, what are you going to do about it?' we said.

Hardly our finest hour.

In 1952, George VI died and his funeral procession went up St James's Street. The directors of Berry Bros. and Rudd kindly asked me whether I would like to watch the procession from their first-floor balcony, and bring a friend. I was delighted to accept and brought along a friend, Philip Bradshaw. I still have the tickets they issued.

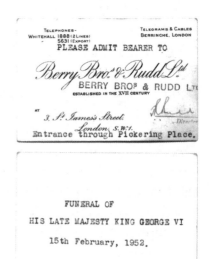

Admission ticket from Berry Bros & Rudd Ltd for watching King George VI funeral from their shop premises at 3, St James Street, SW1

They gave us a splendid meal and, as you would expect from the Queen's wine merchants, lashings of excellent wine. We went out onto the balcony as the procession started to go up St James's Street from Pall Mall to the tune of the Dead March and with muffled drums. The sheer solemnity of it all became too much for the intoxicated Philip and me, and to my lasting shame I have to relate that we burst out giggling. We simply could not stop. We tried everything. We avoided each other's glances and stood apart but there was no way we could stop. The guffaws kept bursting out of our lips however tightly we tried to hold them together. The only reason I dare tell this story is because there was absolutely nothing I could do about it. It was by far the most embarrassing moment of my life.

Chapter five

Lazy but Inspired

During my stay at Rawlence & Squarey, I was studying in my spare time for the exams of the Royal Institution of Chartered Surveyors – or meant to be. I took the correspondence course provided by the College of Estate Management which I was meant to study in the evenings or at weekends. I found the subjects wholly irrelevant to the work I was doing at the office and incredibly boring. In any case, I had had a cushy time in the Army for the last year and had quite got out of the habit of studying in my spare time.

'Hadn't you better go and get on with your studies, now, dear?' asked my mother after dinner as I played ping-pong with her one Saturday night. I'd go up to my room sit at my dressing table and take out my study papers: *Law of Fixtures, Dilapidations and Easements*. I would stare at the text and listen to the boom, boom, boom of the double base at the High Pine Club nearby, where my chums would be dancing with pretty girls. What a long way low sounds carry, I thought. No wonder the natives used tom-toms when they were restless and wanted to get together for a party. At the end of a dance, my friends would be taking their partners out into to the rose garden in the moonlight for 'a bit of fresh air', the lucky beggars. I tossed the study paper aside and picked up another one: *Estate Records and Accounts*. God, how boring. I try another: *Landlord and Tenant*. I think I'll just pop up to the Club for a quick drink.

By the end of the year I had done precious little studying and went into the examination room woefully unprepared. In every exam paper, I found most of the questions unanswerable and indeed left the examination room pretty early.

Oh dear! I thought. Eric Strathon will want to know how I got on. I know – I could tell him that I thought I had done rather well and was thus very surprised I had failed: 'It must have been a very close-run thing – touch-and-go you know, Mr Strathon. I probably got 49%.'

Unfortunately, it didn't quite work out like that. What I didn't know was that Eric, Governor of the College of Estate Management, could gain access to my RICS exam results. When I received notice that I had failed my RICS Intermediate exam, Eric asked me to come and see him. I was ushered into his room and was just about to recite my well-rehearsed spiel: 'It must have been a damn close-run thing' when I looked at his face. Something had clearly gone awry.

'Hilpy, when I heard you had failed your Inter, I asked the RICS for your exam results. They do not make good reading. Here are your marks.'

Oh, God, no. Please not, I thought. I do not want to hear this. When I was a little boy I used to shut my eyes thereby believing I had made myself invisible. If only I believed that now.

'*Landlord and Tenant:* eight per cent...' and so it went on.

'Your father was a Prizeman, you know, Hilpy. That meant he came top of all the candidates. I'm glad he can't see your results. He'd be a bit disappointed.'

I went home resolving to do better. I know, I thought. I will attend evening classes at the College. So for the next year, after work, four days a week, Monday to Thursday, I attended classes in St Albans Grove, Kensington, and got home at a quarter to ten each night.

Most of the lectures I again found boring, but there was one great exception – those given by Hector Wilks. He was the son of Mark Wilks who was junior partner to my uncle, Sir Herbert Trustram Eve. Mark Wilks had died on the same day as my father and their obituaries in the professional journals appeared on the same page. (Gerald's is in Appendix 3.)

Hector suffered from a severe stutter and he thought the best way of getting over it was to throw himself in at the deep end and become a lecturer. A very good one he was too – and the occasional stutter made you pay attention again if your mind was wandering – which it hardly ever did, as he was such a good lecturer.

To my utter amazement, the whole of his first lecture on *Rating* comprised a history of the Eve family, and much of what I have already written about it in these memoirs was gained from him. Apparently, between the wars a small bunch of chartered surveyors, including Herbert and Gerald, used to go the rounds of the various Courts giving expert evidence in appeals concerning rating and compensation.

I got a pretty good idea of what my father used to do in court when he once asked my brother Julian and me to a case he had on at the Law Courts in the Strand in London. I remember the first expert witness, for the Government, was an Army Officer in uniform. It was concerning the value of a large school which was being requisitioned during the war.

Expert witnesses, because of their expertise, are the only ones allowed to give their opinions. Other witnesses are witnesses as to facts only; thus the important thing for an expert is to demonstrate to the court the amount of expertise and experience he has, so that his opinion will be given more weight. Therefore, in an expert's proof of evidence the first thing he sets out in great detail is the extent of his experience. In this case, counsel asked the Army Officer if it was correct that he had valued two or three (not well-known) schools for requisition rental. Yes, he had. And what was his opinion as to the value of this school? He gave a figure.

When he was cross-examined, each time he was asked a question he rustled through a file before eventually giving an answer.

After that, my father entered the witness box, which was near the judge.

'Morning, Eve,' said the bewigged judge.

'Morning, my lord' said Gerald.

I leant over and whispered to Julian:

'This seems rather unfair. Daddy knows the judge.'

Julian agreed.

Gerald's counsel got up and after one or two preliminaries, establishing that Gerald was a Past-President of the Chartered Surveyors Institution, he asked him:

'Is it correct that you have valued the following schools?' He looked down at Gerald's proof of evidence: 'Eton, Harrow, Marlborough, Winchester, Rugby, Wellington, Radley, Stowe, Haileybury, Bradfield, Dulwich, Oundle, Rossall... ' He mentioned a few more.

'It is,' said Gerald.

'And, bearing in mind your unrivalled experience in the valuation of schools, what in your opinion, Mr Eve, is the value of this one?'

Gerald gave a figure and his counsel sat down.

Opposing counsel got up to cross-examine. He asked a number of questions in what I thought was a very nasty way and I soon began to hate the man. Nonetheless, Gerald stood up to him. As each question was put, Gerald looked down at a little sheet of writing paper he had in his hand, obviously crammed with notes and figures, and snapped the answer straight back at him. It was most impressive.

The Court then adjourned for lunch and I went up to Gerald and said what a lot of public schools he had valued.

'Yes', he said, 'and my counsel stupidly didn't mention them all.'

I was just about to say what a nasty man opposing counsel was, when the man himself walked up.

'Eve, I've asked Taylor [the judge] if we can lunch together and he says he has no objection.'

'Splendid' said Gerald. 'I've got a table for us at the Waldorf.'

I couldn't believe it, but that's the way it was with that crowd of expert witnesses and barristers. One day you were on one side and the next you were on the other. It was all a glorious game, with strict rules and complete integrity, and a very profitable game it was too.

Maurice Gibbs, a distinguished surveyor, is recorded in a booklet, *The Expert Witness and The Principles of Evidence* published by the Chartered Auctioneers and Estate Agents' Institute in 1953, as saying:

> I have seen and heard but a few of the giants of the witness box. One or two of them remain firmly printed in my memory. Of the people I have seen I should like to mention one or two. I can recall Sir Herbert Trustram Eve who was what I would call a great witness in the didactic style. He came into the court to tell the court the answer to the question which was bothering them, and there would be a certain impropriety about the fact that someone was going to get up before he had finished and suggest that he might be mistaken. One felt that such a procedure would be quite improper, and that shows the quality of the man.
>
> I recall his brother, Mr Charles Gerald Eve. I would describe him as the omniscient witness. He was calm, he was compelling, he was subtle, and a great witness.

Unfortunately, at the College I found many of the lectures boring and sat at the back, paying little attention, next to a chap called Donaldson who was also following his father into the family firm with evidently little enthusiasm. When we were particularly bored we used to look at pictures of girls in *Paris Hollywood* magazines while others more interested were taking notes. As a result of this incredible stupidity on my part, by the end of the year I had no notes with which to revise. Something had to be done to give me a chance of getting through the exam; I needed to take a lot of time off and have a dirty great swot.

Fortunately, I had stroke of inspiration. I have been called lazy but never stupid. On reflection I *have* been called stupid. The maths master at prep school once wrote in my report:

> 'The fact that periods of brilliance are interspersed with others of almost incredible stupidity may, we hope, herald genius.'

This must have been one of my periods of brilliance. I wrote to the College of Estate Management to say that I had left on a bus a briefcase that contained all my copious, fully-annotated course notes, which I had so painstakingly taken down during lectures. Could they possibly let me have all the year's correspondence course papers? Yes, they could, and I went and collected them all. It cost me a bob or two but that served me right. Rawlence and Squarey then kindly gave me two months off to revise for my exam.

I decided I would work so many hours a week and I then put up a chart on graph paper in my room. I would have a big square for each hour I had to work and would colour it in red crayon when I had worked it. I think the hours were 9 to 11, 11.30 to 1, 2 to 4, 4.30 to 6, and 8 to 10, totalling nine hours a day. At weekends I allotted myself three hours on each day; total for the week: 51.

If I went to the cinema one evening instead of working between 8 and 10 p.m., two deathly white squares would remain on my chart for that evening, for all to see. That meant that I had to make it up at some other time, so that I could fill in the squares retrospectively. I would have to get up early the next day or work late that evening. I divided all the different subjects between the various days and at the end of studying each paper, I went through the questions at the end mentally to see if I could answer them. If not, I went back over the paper to elicit the answer.

The whole system worked magically and I recommend it to anyone who has difficulty in getting down to study in a disciplined fashion.

I usually had great trouble learning *Building Construction* because I thought it the most boring subject of all, but fortunately a new house was being built about two hundred yards away from home; so I used to go there every few days to watch the progress and ask the foreman questions. It proved to be invaluable. For instance, the day before the exam I asked the foreman what those bits of wood were for, which were stuck between the wooden joists on the first floor.

'That's herringbone strutting which is placed between the joists to prevent them from warping', he said.

Blow me, the very next day the first exam question was:

'What is herringbone strutting and what is its purpose?'

Easy peasy.

The one thing I could never quite get the hang of in *Estate Records and Accounts* was how to compile an account. The day of that exam paper I got on the train at Weybridge and found myself sitting opposite a friend of mine studying to be a chartered accountant. I told him of my difficulty.

'That's easily answered', he said. 'If someone owes you some money, you stick the sum on one side [I forget which] of the sheet, and if you owe it to someone you stick it on the other. Then for a balance sheet, for some completely unknown reason, you stick them all on precisely the opposite sides.' I went into the exam with this one maxim foremost in my mind and had no difficulty at all answering any question they could throw at me. I thought: Why the hell didn't someone tell me this before?

My exam technique, my having a somewhat photographic memory, was to swot up the relevant paper till the very last minute before I went into the examination hall. I then wrote down everything I knew, and then decided which of the exam questions most closely coincided with the various pieces I had written.

In the event, I passed my RICS Intermediate exam –with a little help from my friend, one of the invigilators. I had never met him before but a friend in need is a friend indeed. Halfway though the exam I was dying for a smoke – I was a heavy smoker then – and asked to leave the room. I was accompanied by this invigilator and I offered him a cigarette which he accepted with alacrity. I had the exam paper with me and told him I thought one of the questions was ambiguous.

'Let's have a look', he said. I passed him the paper; he studied it and told me what he thought it meant. Jolly nice of him, and perhaps a little irregular, but, after all, if the chap setting the paper can't write plain English, surely you are allowed a little help. In any case, it's too late to fail me after all this time, but I haven't felt safe to tell this story until now. As for the invigilator, he was no youngster and I expect by now he has been out of harm's way for a long while.

I have yet to receive any notification that I came top like my father did. Perhaps the letter got lost in the post. Mind you, my father may have been a Prizeman but it is relevant here to quote from the opening words of his Presidential Address to the Chartered Surveyors Institution on 14 November 1932:

> 'I am the first Prizeman of the Institution who has reached
> this position but I have always been satisfied in my own mind
> that the examiner in question had but a nodding
> acquaintance with the subject.'

Chapter six

Bowlers and Bedbugs

Delightful as my trips into the countryside with Mike Rawlence were, the real eye-opener to me was the management of the urban estates with which the firm was entrusted. One of these estates was in Camden Town where the 99-year leases of the houses were coming to an end. The tenants had covenants in their leases that they must hand over the property in good repair when the term of years had expired; it was our job to pre-pare schedules of dilapidations and serve them on the tenants shortly before the end of the lease.

Some of the tenants had recently bought the tail end of the lease and had no idea that they would have to pay hundreds of pounds to put the property in good repair. Sometimes the front elevation needed rebuilding which would cost about a thousand pounds. The only reward they would get when they had done the repairs would be to be chucked out when the lease expired in a few months' time. It was heartbreaking to have to explain to the tenants what you were doing (making a list of things that needed repairing), what you were going to do (serve a notice on them requiring them to do all the repairs listed), and what you were going to do after that (serve them a notice to quit). Some of the leaseholders assigned their leases to a firm of estate agents who said they were bankrupt. This complicated matters somewhat as we then had to try to trace the previous lessees – who were liable to have disappeared.

The houses themselves were in appalling condition and mostly occupied by tenants living in a state of squalor. I thought at first that this was because the tenants were poor, but I later discovered that the very poorest of people often kept their houses spotless. The worst tenants were those that kept cats and didn't let them out to 'do their jobs'. The stench was indescribable and we often had to spend some hours in the house.

Duncan Rawlence gave me a tip as to what I should do when inspecting the interior of the houses in Camden Town:

'Eve, when you're inspecting houses, always keep your bowler hat on. Bedbugs can't climb up the iron legs of a bedstead, so, instead, they climb up the wall and drop from the ceiling. When you get out of the house just sweep the bugs out of the brim of your hat.'

I adopted this policy rigidly but it proved a little embarrassing once when I was valuing a very posh residence in St John's Wood. The lady of the house let me in and as usual I walked into the drawing room without removing my hat.

'Young man, have you no manners? Did no one ever teach you to take your hat off when you enter a house?' she said. I could hardly reply that I was keeping it on to catch the bedbugs dropping from her ceiling.

The largest urban estate we managed was in Stepney. Although it was extensive, it was almost impossible to make any money out of it. The Rent Restriction Acts prevented us from charging more than a few shillings a week for the rent of a house and the local Council's Sanitary Inspector spent his life serving Sanitary Notices of Disrepair on us as agents. The Notices seemed to me to be mostly concerning broken lavatory pans. How can you break a lavatory pan? I have never broken one in my life and I do not know of anyone else who has ever broken one, but on this estate it seemed to be a daily occurrence and they were expensive to replace.

Looking back on it now, being older and wiser, I think there was probably only one broken lavatory pan on the whole estate. A perfectly sound lavatory pan in a house would be taken out and the broken one temporarily installed, so that a Sanitary Notice could be served. As soon as we had put in a new pan, the old, perfectly sound one was replaced, the broken one put in another house, and the new one flogged off; the proceeds were split fifty-fifty between the Sanitary Inspector and the tenant.

On one occasion, I found two old ladies waiting for me as I arrived back at the office after lunch on a hot summer's day. They lived on the Browne Estate and complained that a Sikh who lived in the house next door sunbathed nude in his garden. I decided to investigate personally.

On my arrival, I asked:

'From which room can you see this unclothed gentleman?'

'From the kitchen,' they chorused. I went through to the kitchen and looked out of the window; I could see nothing.

'No, no, not from that window,' they said. 'From that high one over there. To see him, you'll need to stand on the chair we keep under it.'

Another tenant lived in one of the three-storey houses on the Browne Estate that had a different tenant on each floor. She and her family lived in rooms on the first floor and complained that none of the other tenants ever 'cleaned their public parts' though they kept bringing dirt in on their boots. I was glad to hear, by implication, that there were at least some parts that the other tenants cleaned, and I thought I had better visit the place. On my arrival, I indeed found the hall, passage and stairs were filthy.

I went back to the office and thought I had the ideal solution. I would have a roster. Each week, a different tenant would be responsible for cleaning the public parts. I drew up a schedule naming the tenant responsible for cleaning the public parts for each of the fifty-two weeks of the year, and sent it to all three tenants. Not a problem.

Oh yes? I had a rude and indignant letter from the ground floor tenant saying that in no way would she clean the staircases because she didn't use them. The first floor tenant then rang up to say, using pretty foul language, that she wasn't 'bloody well going to clean the second floor staircase' because she never used it, but there were a lot of gentlemen that did and the tenant above her was 'no better than she should be'.

Oh dear! I had better think of Plan B.

I racked my brains. Eureka! Why on earth did I not think of it before? Each tenant would be responsible for one floor and the staircase leading to it. Simple – and the beauty of this ingenious system was that I would not need to draw up a roster at all. Piece of cake. Excitedly, I dashed off a letter to all three tenants announcing my brilliant solution to this hitherto intractable problem which would, at a stroke, eliminate the cause of so much aggravation.

Heigh ho. Things don't always turn out how you would like them to. The telephone goes red-hot again. It is the first floor tenant:

'I am not bloody well cleaning the first floor staircase every week when the tenant above me messes it up half the time – in fact most of the time, with all her gentlemen friends traipsing in and out all times of the day and night.'

Then a letter from the ground floor tenant saying in effect that she wasn't going to clean the hall every week when she only used it a third of the time. I have omitted the rude adjectives.

Mr Cooper in the General Office looked at the letter and said 'Tut, tut. Such language. Oh, well, Mr Eve. You know what they say. These things are sent to try us.'

Yes, I thought – and they'd better not come in threes.

My third roster, although I say it myself, was a real winner in that it took into account fully all the objections and representations made by all three tenants. To give you a taste of it, here is the preamble:

*The top tenant will clean the second floor every week, the first
floor every second week and the ground floor every third week.*

*The first floor tenant will clean the first floor every second week
and the ground floor every third week.*

*The ground floor tenant will clean the ground floor every third
week.*

Underneath, I drew up the weekly roster with an elaborate
series of boxes designed to make the whole system crystal clear
to even the simplest of folk in our richly varied society. I sent a
copy of it to each tenant and, for good measure, posted it up in
the hall. No angry letter or telephone call followed. It was clear
that, due to my persistence, diplomacy and ingenuity, I had
cracked it.

A few weeks later, I visited the house to see how the system
was working. From top to bottom, all the 'public parts' were
absolutely filthy.

Cooper was right: these things really do come in threes.

The rent collector for the Browne Estate, like every sergeant
major I ever met, and I met quite a few during my three years
in the Army, was a formidable man. Being a rent collector
meant that Sergeant Major Gray visited every house every
week, for all the tenancies were weekly. The 'weekly' part
applied only to the payment of rent, such as it was, for the days
when you could give a tenant a week's notice to quit were long
gone. Gray was our eyes and ears and knew everything that was
going on.

Once a week, Gray would come into the office and make a
report. This week he told us that one of the houses was being
used as a brothel. The madam was the lady on the second floor
of the house with the cleaning roster difficulties.

'Good Lord,' said Mike. 'We can't have that on our estate.
What would Mr Browne say? They might charge him with living on immoral earnings by taking rent from a brothel. We'll
serve a notice to quit on her on the grounds that she is

breaching the terms of her tenancy by using the premises for immoral purposes'.

'With great respect, sir, I wouldn't advise that,' said Gray. 'She'd sue you for libel – defamation of character. No, sir. What you have to do is to take her to court and prove to the judge she is using the house as a brothel. That means you have to produce a witness who has actually been to the house and had sex with one of the girls in payment for money. You'd have to send someone there, sir.'

Mike looked round at me.

'Don't look at me, Mike, I'm not going. Not on your Nelly!'

For all I know the brothel's still there today. I wonder what the staircase looks like. I know it will be well-worn but is it still filthy?

Our sergeant major told us that the police didn't regard bookmaking and brothel-keeping as 'crimes'. They were merely against the law in this particular country, so they had no compunction in accepting bribes from those people in exchange for not taking them to court. Indeed, I was told that every Sunday the Chief Constable, or Chief Superintendent, used to ride down the middle of Commercial Road, Stepney, on his horse to give bookies and pimps their opportunity of handing over their protection money.

One day, Gray came in and said that the widow who lived alone in one of the smaller houses had died and the house was therefore empty. He proposed that Mrs Goldstein move in. The street concerned, like most of Stepney in those days, was occupied by a Jewish community.

'We're not having anyone move in,' said Mike. 'We only get a few shillings a week rent from it which barely covers the cost of repairs. No, we're going to sell it with vacant possession and make a bit of decent money for our client, for a change.'

'Oh, I wouldn't advise that, sir,' said Gray. 'You see, the street has decided that Mrs Goldstein's house is a bit crowded, now that her son's family living with her have had a third child, and that she should move out into number 36.'

'I don't give a damn what the street says, Sergeant Major. We're running this show and we are going to sell this house with vacant possession. Hilary, perhaps you'd be kind enough to go and get out particulars for it, so that we can put it on the market.'

'The street won't like it, sir,' said Gray, shaking his head as he went out of the office.

'The street can bloody well lump it,' said Mike to me after Gray had closed the door behind him. 'The trouble with that chap is that he gets too close to them. We're trying to run an estate, not a Benevolent Fund.'

Six months later, in spite of my getting out a magnificent set of particulars, Mike said to Gray: 'Sergeant Major, why hasn't that house, number 36, sold yet?'

'Because no one could live in it, sir.'

'Why not? It is a jolly nice house, very reasonably priced.'

'Because no one would speak to the new owners, sir. They might also find their newly delivered milk bottles outside the door had somehow fallen down and got broken each morning.'

'Why is that, Gray?'

'Because, sir,' said Gray patiently 'the street has decided that Mrs Goldstein should have the house. Once they've decided that, anyone else who went to live there would find life absolutely intolerable.'

'All right,' said Mike, 'let the damn thing to Mrs Bechstein; with the whole house to herself she'll certainly have room for her grand piano.'

'Mrs Goldstein' said the sergeant major, patient as ever. 'Very good, sir 'and he shot out of the door, giving me a little conspiratorial wink on the way. I hope she enjoyed her time at number 36.

I was very impressed with the community spirit that had grown up during the eighty or ninety years that the estate had existed. No one went hungry, no one went without – much – and certainly no one was ever lonely. When councils decided to clear large areas of terraced housing and either send the people out to New Towns, such as Hatfield, Stevenage and Crawley, or rehouse them in tower blocks, I was absolutely appalled – and have been ever since. The configuration of a housing estate has a fundamental effect on the quality of community life, but unfortunately very few people realise it. They have not had my experience. The community spirit in these high-density terraced estates was superb.

Obviously, as the urban housing estate in this country got to the end of its useful life, it had to be replaced, and some of the housing in slum areas was back to back and lacked proper sanitary facilities. However, there were a number of ways of replacing and updating these estates without wrecking the close-knit community structure that had taken some eighty or ninety years to build up.

For instance, one could demolish one block having slowly rehoused the people in vacant houses throughout the estate. It could then be rebuilt as a modern terrace and the people opposite would then move into it. This procedure could be repeated throughout the estate, so that the whole of it was renewed, causing the minimum of disturbance to the community structure.

Another of my ideas was inspired by a method they use in Sweden; there they produce in a factory what they call the 'heart' of a house. It is transported on a trailer and has standard dimensions so that builders know exactly how to build their site. On one side of the heart is a kitchen with all modern appliances, and on the other side is a bathroom. At either end is a lavatory with a WC and washbasin. This heart is transported to the site, lifted up by crane, plonked in position and 'plugged in'.

On a London estate, one could demolish all the rear gardens and rear additions and drop in on the back a bathroom, a kitchen, upstairs and downstairs lavatories and plug them all in. In individual cases the clip-on rear addition could be delivered by helicopter.

Later on in my professional career, I used to get very incensed with the insensitivity and ignorance of those town planners who were in love with the idea of replacing terraced housing estates with tower blocks. If only they had lived on these estates, or even spent time walking round them talking to the residents. Ignoring the people who were killed, injured or rendered homeless by the London blitz, the estates were improved by the number of bombsites that appeared. These were wonderful playgrounds for the children. The traditional picture I have in my mind is of a mum sitting on a chair by her front doorstep, breast-feeding her child, keeping an eye on her children playing on the bombsite opposite, and having a chat with the neighbours as they pass by. This scene is about right, if the weather was good enough.

It is so sad when ignorant, arrogant planners wreck this intricate, long-standing structure.

One famous architect reacted to criticism of his Council tower blocks by, as an experiment, living on the top floor of one of them. A bachelor, he pronounced it as a splendid way of life, but he did not have the problem of not being able to put his baby in a pram in the back garden, or keeping a weather eye on his kids as they played.

I remember thinking of starting a campaign called (it was the time of The Man from U.N.C.L.E.) Operation BLOCK-BUSTER which stood for *Banning Large 'Orrible Concrete Kremlin-like Blocks Unsuited to the Sort of Tenant Expected to Reside in them.*

I also thought of asking some Royal person to institute the *Royal Society for Laying Tower Blocks on their Sides.*

Years later, I recall one of our Scottish local authority clients boasting that his council had built thirty-five-storey blocks of flats to rehouse people living in terraced houses.

'Is that a good thing?' was the mildest remark I could muster but it wasn't very well received. He thought it was a source of pride. I thought it absolutely appalling.

Another aspect in this matter was that the New Towns were not the panacea many planners and politicians hoped they would be:

First, the inhabitants were not meant to commute to work from their new homes – they do, for the towns, like Stevenage, Crawley and Hatfield, are not far enough away from London; people even commute to London from Milton Keynes which is much further away.

Second, the towns were built much too diffusely. Instead of building terrace houses for close-knit communities, they spaced out all the housing. The result was that no one got to know anybody and communities were not created. People suffered from loneliness. It was called 'New Townitis'.

At the end of my third year at Rawlence and Squarey, Uncle Duncan having retired, Mike Rawlence was in a position to be generous towards me, and gave me the £300 bonus I mentioned earlier for the last year's work. I was delighted. I had also during this year passed my RICS Final examination, but I have no recall how I achieved this. Brigadier Killick, Secretary of the Royal Institution of Chartered Surveyors wrote on 5th November 1952 to tell me I had been elected a Professional Associate:

> ...It gives me particular pleasure to tell you this because, as I told your brother on his election last year, the name of Eve is one to conjure with in the Institution, and I am delighted to think yet another member of so famous a family in the surveying world is a fully fledged member.
>
> With every good wish for a happy and successful career...

THE ROYAL INSTITUTION OF CHARTERED SURVEYORS

Telephone
WHITEHALL 5322

Telegrams
Surveyable, Parl, London

12, GREAT GEORGE STREET,

WESTMINSTER, S.W. 1

AHK/JD

5th November,
1952

Dear Mr. Eve,

I am very glad to be able to report to you that, as a result of the scrutiny of the ballot this afternoon, you are now an elected Professional Associate of the Institution.

It gives me particular pleasure to tell you this because, as I told your brother on his election last year, the name of Eve is one to conjure with in the Institution, and I am delighted to think that yet another member of so famous a family in the surveying world is now a fully fledged member.

With every good wish for a happy and successful career,

Yours sincerely,

A.H. Killick

Secretary

H.M. Eve Esq.,
   C/o Messrs. Rawlence and Squarey,
      6, Ashley Place,
         Victoria Street,
            London, S.W. 1.

**Letter dated 5th November 1952 from
AH Killick, Secretary, RICS**

A charming letter, which I have kept, but I felt it strangely unnerving. It seemed that so much was expected of me and I had so little confidence that I could live up to those expectations. However, my articles were over and it was time to return to the fold.

I went to see Eric (see Appendix 1 for obituary) who said the firm had decided to send me to their Cardiff Office.

However, before I went, Mike organised a very entertaining day out which I will tell you about in the next chapter.

Chapter seven

A Day at the Races

Mike, with Irish blood in his veins, loved horses and was fond of a good flutter. So he decided on one occasion that we should take a day off and go to the races. There were certain essential preliminaries to be gone through, most important of which was to concoct a good story for Uncle Duncan, who, first, had never been to the races, second, would certainly not approve of gambling, and, most of all, would definitely not have approved of Mike and I going to the races in office hours.

Mike took care of this aspect by spinning a yarn to Duncan that we were going down to see someone who might want us to look after his estate for him.

It was the first race meeting at the racecourse at Newbury, Berkshire, to be held since the war. Mike and I drove down with a farm manager, Paddy O'Malley, who was a friend of Walter Nightingale, a very well-known trainer. Paddy said the drill was this: We would be in a different enclosure from him and Walter, so five minutes before each race we would meet at the fence dividing the two enclosures and he would give us any information he might have.

We arrived just too late to bet on the first race, so we went to the paddock to see the parade of the horses in the second race. We decide which horse we fancied and each put on a bet with the Tote. Never much of a gambler, I coughed up two shillings (10p) which I think was the minimum bet in those days. We then went to the boundary fence to meet Paddy. He told us to bet on a certain horse and to put it 'on the nose' – i. e. to win only. We did so and were delighted to see the horse romp home, the two horses we had selected coming a few hundred lengths behind. Was this a bit of luck or did Paddy really know something we didn't?

At the next race, we forbore to choose ourselves the horses to bet on by how they looked in the parade and whether or not we fancied their names. We would wait on Paddy's advice. He gave us the name of the horse which again he said should be backed to win only – not each way. (I later learned that some jockeys are instructed to win but that, if near the end of the race they find they cannot, they are told to avoid being placed, as it could affect their handicap in future races.) Coming into the home straight, Paddy's horse was very poorly placed, so I thought, but then as if by magic (or perhaps design) he overtook every horse in sight and won by half a length.

Now Mike and I were really impressed – and richer; I, only slightly, but Mike quite substantially as he had not been stinting on his wagers. So we were a little disappointed when Paddy told us that the next race was a very open one, but his information was that a certain horse would finish in the first three. On went our money on the tote – a bit more this time – for a place only. The horse led most of the way but failed to stay and finished in third place. Mike and I were now familiar faces as we went to our window at the Tote to collect our winnings from the kind lady for the third time running.

At the next race, Paddy said it was a large field and a very open race. Walter had no horses running in it. He advised us not to bet but to save our stakes for the last race for which he had the best tip of the day, but he wouldn't tell us the name till a few minutes before 'the off'. We did as Paddy suggested and kept our powder dry.

As the horses joined the parade for the last race, Mike and I had a small bet with each other as to which horse would be Paddy's tip. We went to the fence and waited and waited for Paddy. There was no sign of him. How awful, we thought if he never turns up and we lose making our best bet of the day. Then just before the start he hurried through the crowd and lent over the fence.

'This is as good a tip as you've had all day, Mike, m' boy. Nothing is certain in racing but you-know-who has a horse in this race and he has every confidence in him winning.'

'Yes, yes,' said Mike impatiently, but what's its bloody name, Paddy?'

'Come a bit nearer, Mike,' he beckoned. He cupped his hand to his mouth and whispered something in Mike's ear.

'Thanks, Paddy' said Mike and dashed off with me to the Tote. The horses were circling round behind the start so it was going to be a close-run thing to get our bets on in time.

I had already counted all my winnings and decided to put them all on the last race.

Mike rushed up to our lucky Tote window. A voice on the Tannoy said:

'They're under starters orders.'

'Number one to win, please' said Mike pushing forward a bundle of notes. Number one, I will always remember, was called The Leader.

I went to the window with my collection of coins and put on my bet. I noticed the odds were two to one, so for me it was a question of treble or quits. Mike said:

'Hilary, I don't think I can face watching this race. I am going to sit in the car in the car park. After it's all over you can come and tell me the result.' And off he hurried without another word. I had a horrible feeling that Mike had more money on this race than he could afford to lose – or that his wife, Lorna, would approve of him losing. I began to feel rather anxious as the horses cantered up the course in front of the grandstand making their way to the start. I wasn't that worried about myself as I had nothing to lose but my winnings.

Soon the Tannoy crackled: 'They're off!'

Listening to the commentary made me wonder why on earth this particular horse on whom so much now depended had been called The Leader; not only was it not leading but it was not even 'mentioned in despatches'. After a while I spotted the jockey's colours. This was not surprising because there was no one alongside him. How could there be when he was clearly last? However, soon he moved up into the centre of the crowd of horses and got his first mention by the commentator:

'The Leader is coming up on the outside and is making a strong challenge'.

The Leader finally proved worthy of his name, for he took the lead down the home straight, and, more importantly, continued to lead as he passed the winning post.

My first thought was to go and collect my winnings but then I suddenly realised that Mike would not know the result of the race and would be sweating it out in the car hoping that all the cheering meant the favourite had won. I must put him out of his uncertainty as soon as I could, so I turned on my heel and jostled my way through the crowd to get to Mike in the car park.

Mike saw me coming and I put up both hands with my thumbs up. I have never seen a quicker transformation in a man's face than at the moment he saw my thumbs go up. He took in a deep breath and then exhaled, his cheeks bulging.

'Phew! Thank God for that. How much did it win by?'

'Oh, about three lengths eventually, but it took so long coming into the lead that it had me worried stiff.'

'Hilary, I didn't tell you, because I thought you might be worried but I put £100 on that horse.' (£100 then would be about £2500 today.) 'If I'd lost, Lorna would have been livid.'

'Well, it did win, Mike, so how about our going back to the Tote to collect our winnings, eh? You'll probably need some help carrying your lot.'

Mike was nothing if not a gambler and he went on to make millions in property out of his 'big deals' – and, towards the end of his life, lose the lot in one big deal in property on the French Riviera. He died this year, 2004, aged eighty-four, and it seems appropriate that he had asked that at his funeral service, which I attended, there should be a Reading from Rudyard Kipling's *If*, which includes this passage:

> If you can make one heap of all your winnings
> And risk it on one turn of pitch-and-toss,
> And lose, and start again at your beginnings
> And never breathe a word about your loss;

Part III. *Gerald Eve and Co*

Cardiff

CHARTERED SURVEYOR

May 1951

Chapter eight

Come On 'Quins!

The office at Gerald Eve & Co in Cardiff was at 16 Park Place, almost opposite the Civic Centre buildings, which were described as 'the best civic buildings this side of New Delhi'. I thought them very impressive, so the civic buildings in New Delhi – or perhaps the other side of New Delhi, wherever that is – must be truly magnificent.

Park Place was, and I hope still is, a delightful row of early Victorian villas. The ground floor of number 16 was devoted to office purposes, plus a kitchen and dining room; the first and second floors were living accommodation beautifully furnished. The staff included the highly capable Mrs Dolman, who ran the office, apart from the surveying work, and Derek Wade, a chartered surveyor.

In charge was Geoffrey Powell, son of the senior partner of

Geoffrey Powell 1978

the firm, Jack Powell. Geoffrey was the same age as me and had joined the London office a week after me in 1948. We had worked in the basement together colouring plans and developing and printing photographs. Like me, he had come to the London office straight from the Army; he was commissioned in the Welsh Guards, and, like me, on demobilisation, had been articled out – to an agricultural surveying practice in the West Country.

Again like me, he had failed to pass his RICS Intermediate exam. His chances of success were not helped by the fact that he failed even to apply to take the exam. Thus his failure was put down to mere forgetfulness but mine to lack of application. I wish he had suggested the idea to me, but I suppose that for us both to forget to apply to take the exam would smack of conspiracy rather than carelessness.

Geoffrey lived on the premises and I joined him there. He was much more mature than I was, had boundless energy, great capability, enormous charm and was excellent at dealing with clients. He seemed to know much more about surveying than I did. However, he was somewhat temperamental, but when he was on form and in a good mood, no one I have ever met was more fun to be with. There was literally never a dull moment from the time I arrived in Cardiff, to when I left some three years later.

In those days, the cost of entertaining clients could be set against tax and so we seemed to have a limitless supply of drink and cigarettes. Jack Powell, a Welshman, said 'Heavy industry, heavy drinking'. The principal industries in South Wales in those days were coal and steel, and the Welsh certainly lived up to Jack Powell's maxim. Geoffrey and I for our part did our best to help them.

One of our favourite places to entertain clients was the restaurant of the Windsor Arms in the Cardiff docks area known as Tiger Bay. One evening, we were entertaining there the agent of a very large estate for whom we acted, who had driven us there in his own car. I will call him Hugh. On the way back – it was very late at night and no one was about – Hugh, much the worse for wear, came across an avenue of trees on a very wide paved area. He decided to do a 'slalom run' and weave in and out of this row of trees. Remarkably, he managed to negotiate the whole avenue successfully and turned round to do it again. He then decided that one of the trees was getting in his way and hooted at it.

'Get out of the bloody way!' said Hugh, but, not surprisingly, the tree obstinately refused to move. This order to the tree, and the hooting, was repeated once or twice until eventually we did see something moving: not the tree itself, but a police car. Its driver got out and Geoffrey, like a shot, got out of the passenger seat and went over to him. He realised that once our driver got out of the car and started staggering around and possibly swearing at the policeman, all would be lost. Not only that but he knew that if Hugh's offence came to light he would lose his job – and he had a wife and three children to support.

Geoffrey seemed to go on talking to this policeman for ages and ages, but at last the policeman returned to his car and Geoffrey came back to us, saying:

'Of course, Sergeant Williams, I will drive him home'. Geoff got into the driver's seat and pushed Hugh along the bench seat to the passenger side. Geoffrey reversed away from the tree that had refused to move and drove off. Hugh, now oblivious to all that had happened, fell asleep at once.

'How the hell did you manage that, Geoffrey?' I asked.

'Oh, we talked about old times together. He was my platoon sergeant in the Welsh Guards.'

As young bachelors, we were not averse to entertaining young ladies on the premises, especially as we had access to limitless free drink and cigarettes – practically everyone smoked in those days and we were no exception. We decided to ask Carrie and Chloe to Sunday lunch, followed by bridge. These two girls lived in Newport, were cousins, both very pretty, and enormous fun.

On Sunday morning, Geoffrey and I had our usual pre-prandial tumbler of Dry Sack sherry and I said I'd go off and collect the girls from Newport, about twenty minutes' drive, while Geoffrey got the lunch ready. When I got to Newport, the girls weren't quite ready, as girls sometimes aren't, and there was some delay while they 'got themselves up'; thus it was over an hour before we returned home.

I opened the front door and called:

'Geoffers, yoo-hoo, we're here. Sorry we're late.'

There was no reply. I walked up to the staircase and there at the bottom were a few Brussels sprouts. A third of the way up the stairs were some roast potatoes. As I ascended the staircase I was met by, in succession, roast parsnips, an upturned gravy boat in a sea of gravy, an upturned jug of mint sauce and, on the top stair, a leg of Welsh lamb. Finally, along the passage was our chef de cuisine himself, sleeping peacefully and hugging to his breast affectionately an empty wooden tray.

On another Sunday morning, Carrie rang up to say that their boiler had burst and could she come and have a bath at our house.

'Delighted,' says Geoffrey. He went and took the key out of the bathroom door and, when Carrie arrived, told her the key had been lost for ages; but not to worry – we both knew she was in there. Of course, as soon as she got in the bath to have a good soak, every few seconds one or other of us went in the bathroom on some pretext or other to get something, while she desperately tried to cover up those parts of her anatomy which we were not meant to see. She spent a very long time lying back in that bath so, on reflection, I don't think she minded too much.

Geoffrey on one occasion was having a bath when some thought, I know not what, provoked his extreme annoyance, and to give vent to his anger he threw the metal bath-rack out of the window. The next morning, Monday, Mrs Dolman entered the front garden and saw the bath-rack. She picked it up and on entering the office asked me how it had got there. I explained and when Geoffrey next had a bath, he found there was some pink tape attached to the bath-rack, the other end being firmly attached to one of the taps. A practical woman, Mrs Dolman.

That wasn't the only thing that Geoffrey threw. On one occasion in the kitchen one weekend, I managed to cause Geoffrey extreme annoyance – one cook per kitchen is plenty – and he threw a piece of china at me. It missed. He then threw a further twenty-one pieces at me, all of which missed. I don't think he was trying very hard to hit me.

It was a Saturday night after the shops had shut and we felt it would be a little embarrassing if there were no coffee cups and saucers for the staff on Monday morning. Fortunately there was a fair in Cardiff at the time and off we went to see if we could win a tea service as a prize. We spent a fortune rifle shooting and taking part in a large number of other competitions but failed to win the right to claim any sort of prize. Eventually, Geoffrey very sensibly negotiated with one of the stallholders to buy a tea service at, of necessity, an outrageously high price. As the one who had caused the annoyance and the resultant breakages – I did egg him on a bit – I felt obliged to contribute half the cost.

Some years later, after I had returned to Gerald Eve's London Office, I got a call from Carrie and Chloe. They were going off skiing and returning by train to Victoria Station in the evening. Did I know of anyone who could put them up for the night before they returned to Newport the next morning? I rang an old friend of mine, John Murray. He explained he had only one bed, admittedly a double one, but after hearing from me what lovely girls they were, said he 'would be prepared to put himself out to accommodate them'. I said we could collect them from Victoria together in my car and I would then find out if, having met him, they wanted to accept his kind offer.

I rang up the girls and explained the situation: there was only one bed; there wasn't even a sofa. We decided that once they had met John, they would let me know whether they fancied him sufficiently to accept his offer. If they didn't fancy him –

and in my experience girls always did – they would say to me in the car: 'What's the time, Hilary?'

On the appointed day, I picked up John and went and collected the girls from the station. On my way back I thought the girls, chattering away to John, might have forgotten the tactful arrangement we had made concerning their willingness or otherwise to accept the somewhat cramped accommodation being offered them.

I turned to the girls and said: 'Would you like to know what the time is?

'The time, Hilary,' they chorused. 'Who the hell cares about the time!'

Many years later I reminded John of the occasion and asked: 'How did you get on that night?'

'Pretty often, Hilary, as I recall. To be honest, it was the best night of my life. I literally didn't know which way to turn for the best.'

Lucky beggar.

In those days, the early fifties, rugby football was a passion amongst the Welsh. Unlike today, there were thousands of miners, and every mining town in South Wales had its rugger team.

To be in Cardiff on the morning of a Rugby International being played at Cardiff Arms Park was really exciting. To catch a whiff of the excitement, you had to go to one of the pubs in Cardiff where the talk would be about nothing but the forthcoming match. You would not find a more knowledgeable crowd anywhere and, still today, at the Arms Park's successor, the Millennium Stadium, you will get more applause for a good piece of rugby from an opposing team than probably anywhere else in the world.

Shortly after I had arrived at Cardiff, I went to the Arms Park to watch my home team, Harlequins, play Cardiff. I stood amongst a crowd of miners, shouting out in what they must have thought was an alien, posh voice:

'Come on 'Quins. Come on 'Quins.'

A man tapped me on the shoulder:

'Look you, boyo, when you come to Cardiff Arms Park and stand here, you cheer for Cardiff.'

I looked at him, and the others round him. They were all looking at me. I got the hint:

'Oh, all right then. Er… come on Cardiff. Come on, Cardiff.'

I was fortunate enough, during my three years in Cardiff, to watch the match where the Cardiff Rugby Football Club, captained, I think, by the legendary Cliff Morgan at fly half, beat the hitherto all-conquering New Zealand touring team, the All Blacks. It was the only match the All Blacks lost in their tour. Someone reading this will know just when was the last time that the All Blacks were beaten by a club team – once in a blue moon, I should think.

Shortly after I came to Cardiff, Geoffrey thought it would be a good idea to have a party for all our clients at St Mellon's Country Club. It would be an opportunity for our clients to meet the new arrival and, what was more important, to meet the three partners of Gerald Eve and Co who were based at the London Office. The idea was to demonstrate that when we young lads gave advice to our clients we were backed by three very wise and experienced partners in London. All three partners agreed to attend the party, but as time wore on they all found some excuse not to attend and dropped out one by one. So Geoffrey decided to send the invitations out saying: 'Geoffrey Powell and Hilary Eve invite you to a party…'

On the appointed evening, Geoffrey and I stood there to receive our guests and as each one came up to Geoffrey they said: 'Geoffrey, this must be your engagement party. Congratulations. I am dying to meet Hilary, your fiancée. I bet she's gorgeous!'

Their faces fell when Geoffrey said:

'Er, actually I'm not engaged. This is Hilary.'

It was a damn good party, unrestrained as we were by the presence of the three partners. The Welsh really do know how to enjoy themselves at a party, and our guests certainly lived up to Jack Powell's maxim of 'Heavy industry, heavy drinking.'

People used to feel sorry for the Welsh because in those days pubs were not allowed to open on a Sunday. There was no need to, because the Welsh got over the problem by forming numerous clubs. Every Welshman belonged to a club and their licences allowed members and their guests to drink on Sundays. Not a problem.

It is an extraordinary thing, but whereas I can well remember much of the work I undertook at Rawlence and Squarey and later on at Gerald Eve's London office, I cannot recall any of the work I did at the Cardiff office. However, I can call to mind all the fun Geoffrey and I had and all the japes we got up to.

Geoffrey said to me one evening that he had made a date with some very attractive girl for Saturday night but that, unfortunately, he was now unable to attend. He had to go home to Berkshire to see his mother who was not well. Would I be kind enough to stand in for him? As an added incentive, he said she was very beautiful and I would find her adorable.

'Well, all right, then,' I said, always ready to put myself out to help a chap out of a jam – and unfortunately being totally unaware of Geoffrey's fondness for practical jokes.

At the appointed time, I drove up to the address Geoffrey had given me, a somewhat humble dwelling, rang the bell and was greeted by a fat, jolly man in a very loud check suit; he was dressed just like a bookmaker, which wasn't surprising as it turned out that's exactly what he was.

'Ullo, you must be 'Ilary,' he said in what seemed to me to be a genuine cockney accent. 'Edith's just comin' out. Look, 'Ere's a fiver; give 'er a good time.'

Edith appeared. I looked at her. Geoffrey, you are a lying bastard, I thought. The poor girl is as plain as can be and, as we
used to say in those politically incorrect days, 'as common as
dirt'.

When I told her my name was Hilary, she was nonplussed
and said she would call me 'Ray', pronounced Rye. The girl was
pleasant enough but I took good care to spend her father's five
pounds in some venue where I was sure I would not come
across any of my friends.

Geoffrey had seen the girl in a pub and engineered the whole
thing as a practical joke on me. It was many years before I got
my revenge on Geoffrey, but get it I did, with interest – but that
is a story I will tell you much later on.

One of Geoffrey's friends was Lincoln Hallinan who was a local
barrister and a member of the City Corporation. Stephen
Potter with his Lifemanship and One-upmanship books was
our great hero. This coupled with Geoffrey's love of practical
jokes resulted in one or two amusing happenings. Dining with
Lincoln at the County Club one evening Geoffrey was boasting
to him in his cups about his manservant who ran his bath, laid
out his clothes, waited at table, etc.

Geoffrey asked Lincoln to dinner. This posed a bit of a problem because the manservant Geoffrey had been describing did
not exist. Not to worry – there was a gardener, Evans, who used
to come once a week and Geoffrey decided to turn him into our
butler for the night. Evans borrowed a dark suit belonging to a
chauffeur friend of his, and Geoffrey tutored him in the art of
butling.

Lincoln arrived and Evans proved to be an apt pupil, for all
went like clockwork – until it came to the pudding. Lincoln
pointed out he had no spoon and Geoffrey asked Evans to get
one. He was out of the room for quite a while, the reason being
that he decided he could not bring a spoon in his hand to present to Lincoln, but would have to find a tray. He found one

behind the kitchen stove, returned to the dining room and stood beside Lincoln with the tray. Lincoln was in the middle of telling a story and did not want to interrupt it to take a spoon, but Evans thought he had waited long enough; he gave a good hard nudge with the point of his elbow on Lincoln's upper arm; Lincoln turned round, in some surprise and with raised eyebrows, and looked up at Evans who gave a jerk of his head towards the tray. We all looked at the tray: it was a tin one which had never seen the light of day for many a year and was covered in rust, grease and a few cobwebs. Lincoln, the perfect gentleman, took the spoon, cleaned it and polished it with his napkin, and started on his pudding, Geoffrey, most definitely, had not pulled it off.

Lincoln had the last laugh because he asked us back to dinner in a private 'hospitality' room in the City Corporation building. The food and wines were, of course, superb – served impeccably by a real live butler.

I left Cardiff in 1954 to go back to our London Office and heard no more of Lincoln, but the next time I saw him I got a bit of a surprise. It was on the television in 1969 and there was Lincoln, now Sir Lincoln Hallinan, clad in the robes and chain of office of Lord Mayor of Cardiff. He was in a procession through the streets of Cardiff that he had organised to mark the investiture of the Prince of Wales.

Part IV. *Gerald Eve & Co*

Mayfair, London

CHARTERED SURVEYOR

1954

Chapter nine

Put and Take

Back at 6 Queen Street Mayfair, I found the firm had expanded into the house next door, No 5. I was again put in the same room – on the ground floor at the back, but this time I shared it with John Lewin. John was the most highly paid assistant in the office: his salary was £1000 a year. He made an immediate impression on me when he said that he had bought a new type of writing implement called a Biro. He explained that it worked by means of a steel ball at the end which, as you moved it along the paper, revolved, allowing the ink above it to coat the ball. It cost 65/– (£3.25) which was more than a day's pay for the highest-paid chap in the office. I tried it out and, as everyone now knows, it worked a treat. It has also come down in price quite a bit, even ignoring inflation during the last fifty years.

John said I could come and help him reference a factory that morning, which would be his first opportunity to take his Biro out to give it a field trial. To 'reference' a property involved inspecting it and recording in a systematic manner all relevant details in order to make a valuation for a particular purpose. This inspection was for the purpose of making a rating valuation.

Since much of these memoirs are from now on going to be about my rating work, I had better explain briefly what rating is.

Rating was started in Good Queen Bess's time by the enactment of the Poor Law Act of 1601. Every occupier of land or nearly every – there may have been certain exclusions such as churches – was required to pay a tax based on the annual value of his land. This was a sort of income tax really because in those days, before the Industrial Revolution, most people's incomes were, directly or indirectly, derived from the land.

I'll skip the evolution of rating for the next three hundred years or so, in case you find it boring, and come to the Rating and Valuation Act 1925; this provided that a Valuation List for England and Wales must be prepared every five years – it was slightly different in London. These Quinquennial Revaluations had been carried out in 1930 and 1935 but the one due in 1940 was postponed because of the outbreak of war in 1939.

As I explained in Chapter 2 the task of making the rating valuations was taken away from the local Councils by the Local Government Act 1948 and given to the Inland Revenue. The I.R. decided to make a new Valuation List for 1953, but found the task too much and got an Act passed to postpone it till 1956.

This meant that the ratepayers would at a stroke have their assessments increased from 1935 values to those of 1956 – which, due principally to inflation, might be two or three times as much. For the moment, householders were going to be let off lightly – they would be assessed at 1939 values, but the commercial ratepayer was going to be hit hard – even though the rate in the pound payable on the rateable value would obviously go down. Clearly, it was going to be a good time to be a practitioner in rating valuations, something which Gerald Eve had made his name as, and it was certainly going to be a busy one.

The factory we were going to reference was that of His Master's Voice at Hayes, Middlesex, owned by Electrical and Musical Industries (EMI). On arrival, we entered a vast factory filled with people, each of whom seemed to be standing in front of an Aga cooker. I stood and watched one of the men. He was making shellac records, which are nowadays called 'seventy-eights' because they revolved on the gramophone turntable at seventy-eight revolutions per minute. The man picked up something that looked like brown plasticine, lifted up the lid of the 'Aga Cooker' and put the plasticine (it was shellac actually) in the middle of the hot-plate and pulled the lid down. He waited a few seconds, opened the lid and took the now black gramophone record off the hot-plate and put it on the bench beside him. The top of the hot-plate and the bottom of the lid had each got attached to it a circular steel plate to imprint the grooves on to the record which the needle travels over when the record is played on a gramophone. He then stuck a label on each side giving the name of the tune, artiste, etc. which of course included the famous HMV logo of the little white dog looking into the horn of the gramophone to hear his master's voice.

In another area was a production line for portable gramophones. When nearly complete the gramophones were placed on three different conveyor belts and some of the parts were sprayed with different coloured paint and had different coloured fabric put on the outside. I was a little surprised to find at the end of the lines three different gramophones – Odeon, Parlophone and His Master's Voice.

John Lewin led me outside the building to look in the dustbins and soon found a 'pick-up' for his record player that had been discarded because it had some minor scratch.

'Just what I was looking for,' said John, pocketing it.

I found the whole experience quite fascinating and felt sure I was coming into a really interesting job at the London Office. And so it proved over the next thirty years.

★★★

The next job I was given was to go and reference by myself some tinplate works in the Midlands, at places like Stourbridge, Dudley and Brierley Hill. These were owned by Richard Thomas and Baldwins, the Baldwin bit being the family of Stanley Baldwin, Prime Minister three times in the period between the Wars. The firm became part of the British Steel Corporation when a Labour Government nationalised the steel industry.

On behalf of the companies, we were going to appeal against their new rateable values and it would be our job to get them reduced. We were not going to be paid on a results basis, as I am told some clients insist upon these days. There was an agreed fee, in accordance with the Scale of Charges issued by the RICS. The fee was payable regardless of whether we got the assessment reduced or not, but you wouldn't stay in business long if you kept accepting fat fees from clients and failed to make them over the years a saving in rates greater than the fee.

I checked in at a two-star hotel in Stourbridge, had dinner there and retired for the night. In the morning, I went to the reception desk and told the two girls I had fourteen complaints; I handed them a list of them. I thought they might be somewhat concerned, but they kept saying:

'Fourteen complaints, fourteen!' and alternately giggled and roared with laughter. I had hoped to be taken seriously so I checked out and decided to stay somewhere different for the rest of the week.

The towns were, I thought, pretty ghastly. All the buildings were pitch black with soot from the various chimneys, and the Victorian streets seemed to consist of alternately fish-and-chip shops and public houses.

I went to the first property, at Stourbridge, to get cracking on referencing it. There were no plans of it so I had to measure it up. I was armed with a five-foot rod, a clipboard, some printed Gerald Eve Referencing Sheets and a pencil. I will start on the main building, I thought, measure its length, width and height, and then work out the cubic content ready for pricing at so much a cubic foot.

I walked through the small office block at the front and, as I passed through a doorway at the back, was confronted by a truly amazing sight. There, stretching in front of me, as far as the eye could see, was a sea of men and machinery. The noise was deafening. I walked up to one part which comprised a very long, revolving tube that came in one end of the building and went out the other. In olden days it had been made to revolve by the water from the river nearby, for then there was no power to be had except that from wind and water. Attached to this tube, in a long line, was quite a number of what looked to me like large, steel mangles. I watched to see what was happening.

Either side of each 'mangle' was a man stripped to the waste and sweating profusely. By his side was a bottle containing a brown liquid – it was tea – which he swigged quite often to prevent dehydration. Round his neck hung a towel; perhaps it was called a 'sweat-rag'. One man was a 'putter' (which rhymed with footer) and the other was a 'taker'.

The drill was this. Behind the putter was a large furnace. The putter opened the door with a long pair of tongs – the heat radiating from the open door hit me like a brick, even though I was standing at some distance. He then took out a white-hot, rectangular sheet of tinplate and put it through the rollers of the mangle. The taker took it – what else would a taker be there for? – and the two of them played 'put and take' together two or three times until the sheet started turning from white-hot to red-hot. It then went back into the furnace again for a reheat. When it had been thinned out sufficiently, it was put in a pile by the side of the taker.

Fascinated as I was by all this, I recalled that I had entered this building to ascertain its various dimensions. I looked to one side, then to the other, and then looked up. The building was completely amorphous. Essentially, it comprised a number of separate steel roof trusses of varying dimensions and heights; indeed the height of the structure altered every few yards, and the length and width of the building also varied every few yards. How on earth was I going to measure it up? I walked out of the shapeless mass in despair and sat down on some stone steps in the hot sun. If the partners had decided to throw this job at me to stop me thinking I knew it all, they had succeeded, but to be honest, I never thought that – and never have. The beginning of knowledge, it is said, is when you start to realise you know nothing.

After pondering for a while, I decided to take a tip from the 'Horse of the Year Show'. There, before the horse started to jump those horrifically high fences, he was taken over a little low one, about two feet high, to give him confidence. Easy, peasy, I thought. That is what I would do. I looked up and saw a small, oblong brick building with a flat roof. You couldn't get easy-peasier than that, so I got up and measured with my five-foot rod its length, breadth and height. I called it, not surprisingly, building number one, made a note of its construction and, after peeping through the window, described it as a store. I was on my way.

I then traipsed round the whole of the site and referenced every building but the big, amorphous one. Then I went off to lunch and had another think. I decided the best thing to do was to take one measurement each of length, width and height and hope that they were about average. I knew one thing for sure: no one would ever prove me wrong; and no one ever has. It's too late now because all the buildings were demolished ages ago.

At the end of the day, I went and checked in at a new hotel where the receptionist told me to go and have a look at the nearby town of Bewdley. I did and it was quite delightful. For some reason the town seems to have missed the Industrial Revolution and its attendant Victorian housing, apart from a few villas. The result was a charming, small Georgian town nestling by the river. No wonder Stanley Baldwin, on receiving an earldom after he had been Prime Minister for the third time, chose to be known as Earl Baldwin of Bewdley.

★★★

Like any job, I suppose, I did have some embarrassing moments. Once I had to go and value a factory that made rubber goods. During my inspection I came into one vast room where there were scores of young girls sitting at tables. On each table was something wooden sticking up which was called a 'form'. It looked like a trowel handle but I soon realised it was meant to represent something rather different. On each girl's right was a large box of French letters, which people nowadays seem to call condoms. The drill was explained to me:

The girl picked up the condom and unrolled it over the form. She then pressed a pedal which caused air to come out of a hole at the top of the form. This inflated the teat at the top. The girl then put her hand over the top to see if there was any air coming out. If she could not feel any air, she rolled up the condom and put it in a little packet and then into a carton. If she did feel air coming out, then the condom was rolled up and put in a bin by her side.

I was told, most unofficially, that the rejects were sold to a dealer and made their way to Turkey. Whenever I go on holiday to Turkey these days, I no longer wonder why it is more and more crowded each time I go there. The birth rate must be enormous.

Then there was the ballet school in London which I was valuing for rating. The Principal was showing me round the building. She went up to one door, pulled it right back and without looking in said: 'This is the girls' changing room.'

I looked in. The girls were in various stages of nudity. They looked up and giggled at me – too quietly for their Principal to hear. I smiled back at them enjoying the view and pretending to make a note or two on my clipboard.

'I see,' said I, but I never told her what I saw.

★★★

The most alarming moment I had whilst inspecting a property was at the old Covent Garden fruit and vegetable market. Some porters were unloading bananas down a trap door into a banana ripening room where I was standing. As a large bunch of bananas was taken from the trolley with a hook and dumped on the floor there was a shout from one of them:

'Look out!' he said and ran back.

There, just beside the bunch, was a very, very large, black, hairy spider. One of the five porters took up his banana hook, which fortunately was at the end of a long handle and lunged at it. It started to run hither and hither and we all ran further back, but after two or three stabs at it the porter killed it.

I went up to look at it and, since I thought that no one at the office would believe just how big it was, I made a life-sized drawing of it. When I returned to the office my colleagues were all duly impressed.

At the time I didn't know what sort of spider it was, and I am glad I didn't because on consulting *The top Ten of Everything 2003* by Russell Ash – (Dorling Kindersley 2002), under the heading Top Ten Deadliest Spiders I find the entry:

No. 1 – **Banana Spider** *(Phonenutria nigriventer):*
> The banana spider…yields 6mg of venom, with 1 mg being
> the estimated lethal dose for humans.

My God, he could have killed all five of us – and still had one dose to spare for the next comer!

Chapter ten

A Shopping Spree

When you enter a family firm where you are the son of an ex-senior partner you expect to be thrown a broom, at least metaphorically, and told to sweep the floor. I am not sure whether this is because people resent your privileged position, to show you that you've got to do lowly jobs before getting to be a boss, to judge your reaction, or simply to see how good a job you make of it. Perhaps it's a bit of all four. Certainly my tinplate works was a really tough job for a young lad to do solo, but I heard years later that they thought I had done it rather well. I only wish they had said so at the time to give me some encouragement, but perhaps they thought I might need sitting on rather than encouraging. That was not so – at least not then.

The firm's next problem was what to give me for my next job. A number of people were asked if they had any work to give me, and not surprisingly they all had the same idea. They all had a few retail shops where they were instructed to deal with the new rating assessments which would come into force at the national rating revaluation. The new Valuation Lists were to be published in December 1955 and come into force on 1 April 1956 and these people decided to get rid of their shop rating jobs on me.

Each shop had to be measured up very accurately, a note made of its construction and its trading position and the details of any comparable properties. The method of valuing shops for rating was to calculate the areas of the front zone of the shop, then the middle zone and then the rear zone. Mark Wilks, my lecturer in rating at the College of Estate Management, told me zoning was invented by my uncle, Herbert Trustram Eve. He was valuing a row of shops in Bedford which had been converted from a terrace of houses. He valued the front rooms at 1/– (5p) per square foot, the backrooms at half that price, (6d) and the rear additions at a quarter (3d).

The Assessment Committees and the Courts accepted this method and today it is used almost universally for valuing shops for rating purposes. It is known as the Zoning Method and the three or sometimes four zones are known as A, B and C and/or Remainder.

The depth of the zones varied: some, from front to back, were 15 feet, 25, and Remainder, some were all 20 feet zones, and some, like Oxford Street, London, had 30 feet zones. Now I expect they have all gone metric.

The upper and lower floors of the shops were then put at a proportion of the Zone A price, such as one-sixth for first floors and one-tenth for a basement, depending on their quality and ease of access. It is a method of valuing shops at so much per unit of frontage (which is the thing that really matters), making adjustments for varying depths and additional accommodation. The beauty of the zoning system was that once you had decided on the Zone A price all the other areas of the shop were automatically valued, as they were all at a certain fraction of the Zone A price.

So the people who so kindly palmed off their shops on me gave me what they thought was their 'rubbish' – a nuisance involving a lot of work and a difficult job for a small fee. The shops were mostly multiples so each one was in a different place which would entail a lot of travelling, which in turn meant time and expense.

Moreover, it was often very difficult to think of a good reason why your client's shop should have its assessment reduced when all the shops in the parade were 'peas in a pod'. Not only that, the rating assessments of the parade, even if they might have been a bit high when they were made, would be low by the time the appeal was put in, because of inflation. So, very rarely was there a chance of getting the whole parade reduced. You had to prove that yours was worth less than all the rest. You might convince the Inland Revenue Valuation Officer (VO) but, if you couldn't, you had to convince the Local Valuation Court. And if you couldn't convince them, you had to convince the Lands Tribunal, where the loser would, typically, pay both sides' costs. After that, you could appeal to the Court of Appeal and thence to the House of Lords, but these latter two appeals were only allowed to be on points of law – not on value.

Apart from the difficulties mentioned above, you had to do much research into rents being paid for comparable properties, try to get rents from shop agents whose time you were wasting, and liaise with other rating surveyors. At the end of it all, the fee was miserably small – which often was just as well if you failed to get a reduced rateable value for the client.

When the entire firm's 'rubbish' had been given to me, I sorted all the shops into their different locations. There were shoe shops, greengrocers, tobacconists, ironmongers – about two hundred of them in all – and some were quite close to each other. I had to make some sort of plan as to how I was going to reference them, so I bought a very large map of England and Wales and had it mounted on board, framed and put up on the wall. I then bought some pins with coloured bobbles on the end, some red and some blue. I stuck the red pins in the towns in which the shops were situated and I changed each red pin for a blue one once I had referenced it. Where there were a number of shops in the same town, I used a large pin containing a circle of paper on the end on which I could write the number of shops in the town.

I then planned my route for each week and went out and started referencing them. When I came back to the office at the end of the week it was very encouraging to take out the red pins from the towns I had been to and replace them with blue ones.

Sir Charles Clore (1904–1979)

Some time later, Jack Powell ('JP'), Senior Partner, called me up to his office.

'Hilary, they tell me you are the shop rating specialist, so you are just the man for this job. Charlie Clore has just made a successful takeover bid for J Sears (Trueform Boot Co.) Ltd and has instructed us to deal with all the rating assessments of their subsidiary, Freeman Hardy & Willis. Here's the file I've just opened. You'll find they have factories and warehouses in Leicester and Kettering; also one or two shoe shops. Here is a list of them. This job should be right up your street.'

I retired to my room and looked at the list of 'one or two shoe shops'. Right up my street? They were right up every ruddy street. There were 550 of them spread out from Penzance, Cornwall to Alnwick, Northumberland. Wow! Oh, well I thought, I have only got one pair of legs and can only inspect one shop at a time – and with a bit of luck the partners may get someone to help me. Meanwhile, the next thing to do right now is go out and buy four gross of red pins and ditto of blue.

It was quite good fun finding all the towns and sticking the pins in; it improved my knowledge of geography enormously. My map now really looked as if it had got measles. When one of the partners looked at the map he was rather taken aback at the work I had to do; he kindly arranged for our Cardiff office to deal with all shops in Wales and the West Country and for our office in Bootle to deal with those in the north-west of England

That was just as well. Little did I know that Mr Clore had by no means finished with his takeover bids. How it all started is

Douglas Tovey of Healey & Baker

explained by Charles Gordon in his book, *The Two Tycoons. A personal memoir of Charles Clore and Jack Cotton* (Hamish Hamilton 1984):

> Clore, untypically forthcoming, had told Douglas [Tovey of leading shop agents Healey and Baker] that his dearest ambition was to assemble the most formidable portfolio of property in the land.

With this in mind, Tovey mentioned to Clore one evening whilst having drinks at Clore's house in Park Street, Mayfair that he should make a bid for J. Sears, (Trueform Boot Co) Ltd. Clore expressed interest and so the next morning Tovey sent round to him a copy of the Exchange Telegraph card for the company. In his covering note he said he thought the surplus value over the balance sheet figures was £10m. Clore, advised by his solicitor friend, Leonard Sainer wasted not a second. They moved swiftly, stealthily and effectively. A quick, concealed build-up of shares purchased in the market, and within two days they had informed the Stock Exchange of the bid and sent a formal letter round by hand to the Chairman of Freeman Hardy & Willis. Charles Gordon wrote:

> A more galling bolt from the blue had never been received by any board of any British company.

He concluded:

> The purchase of Freeman Hardy & Willis was probably his [Clore's] most significant acquisition in that it provided his group with an enormous equity surplus at an early stage of its growth (in today's values something like £100m.).

Leonard Sainer

91

That was written in 1984, so today that figure could be multiplied several times over. All in all, Clore made a bob or two from buying the company formed by Mr Freeman, Mr Hardy and Mr Willis.

Charles Clore invented the 'hostile' takeover bid, and his Freeman Hardy and Willis coup put the fear of God into boardrooms all over the country. Boards of directors met in near panic and asked themselves questions. Will we be Clore's next target? Are we making the most of our assets? How much are our properties worth? What rate of return are we getting on our assets?

When Clore put in a bid in 1959 for Watneys (five breweries and 3670 pubs), there was outrage at what Clore might do to a national institution. Would he turn our local into a shoe shop? There were sighs of relief all round with the announcement that the bid had been withdrawn 'after a friendly discussion between the two chairmen'. I'll bet it was friendly.

A few days after JP had given me the list of Freeman Hardy & Willis's shoe shops, he summoned me to his office again.

'Ah, Hilary, here's another little lot for you. Should keep you busy for a day or two.' He handed me a little booklet. Included in the J.Sears package were another 250 shops trading as Trueform. The manufacturer and the retailer of red and blue pins were going from strength to strength, for I had to go out and buy another 250 of each colour.

Clore's next move was to sell all the freeholds of Freeman Hardy & Willis and of Trueform in several large tranches, mostly to the Legal and General, at the same time taking back ninety-nine year leases with rent reviews every twenty-five years. With nigh on £10m raised from these sales he financed further takeover bids. Indeed, for the rest of his life, Clore never stopped making takeover bids. I knew when each was made because after each takeover I was given the job of dealing with all their rating assessments. Clore took over Manfield, Dolcis, Saxone, Cable Shoes, Lilley & Skinner, Curtess, Philips

Character Shoes, and a few more I have forgotten, and eventually amassed 2000 shoe shops under the mantle of British Shoe Corporation. It became the largest manufacturer and retailer of shoes in Europe

With such a vast organisation, it became necessary to create a central warehouse, which was at Leicester, and I remember going round it. It was quite amazing – especially for its time. Trucks were moving slowly along the aisles entirely unattended and would suddenly take a turn to the left or right. I think there was something under the solid floor which was directing them.

Later, I came across a chap in a little glass kiosk and asked him what he was doing. He explained that suppliers' lorries were coming in from all over the country unloading goods from the factories, and at the same time delivery vans were coming in to be loaded up with a mixture of shoes to be delivered to their 2000 branches. His job was to get the right shoes on the right lorries. At the moment he had a load of slippers which had just arrived and he could not find anywhere to put them, so he had stacked them at the top and they were going round and round the building until he found somewhere to put them. I noticed he was wearing an RAF tie and asked him what his job was before this.

'I was an air traffic controller at London Airport,' he said. The job seemed pretty similar to me as I watched him while his lorry-load of slippers circled round and round until he found a suitable landing place for them.

But it wasn't just shoe shops that Clore took over. He bought the Furness Shipbuilding Co and Bentley Engineering (the only two of his properties where my firm did not deal with the rating assessments). He went for the big stores as well. He bought Selfridges with Lewis's which had vast department stores in the Midlands and North. He lumped all these together with a few property, development and building companies, etc. and called them Sears Holdings.

Since there were national rating revaluations in 1956, 1963 and 1973 my firm did a mass of work for Clore over the years. I remember just before the 1963 revaluation his property director, Aubrey Hawkins, asked me why he should instruct Gerald Eve & Co to do all their rating rather than some other firm. I told him the reason was that we would overall get bigger reductions in their rates bill than any other firm. Well, you may think: you would say that wouldn't you? Yes I would, but I honestly believed, and still do, that by then we had got shop rating down to a fine art, second to none. But by the time you have finished reading these memoirs you can make up your own mind. Certainly our clients seemed to think so, because by the time we had finished doing the 1973 revaluation work we had received instructions to deal with some 7,000 shop assessments. There were chains of tobacconists, grocers, fishmongers, greengrocers, stationers, ironmongers, jewellers, supermarkets, betting shops, ladies fashions, men's fashions, sport shops, shoe shops, newsagents and off-licences, all with their attendant Head Offices, factories and warehouses. My partner, Michael Hopper, with whom I shared the fourth floor of our six-storey London Office in Savile Row, received instructions on larger retail units: the Debenham Group (126 department stores), the C & A stores and some 250 Marks and Spencers stores.

However, I told Mr Hawkins I would give him some facts and figures for his directors to ponder over. I sent him a schedule showing in detail what percentage reductions we got for each subsidiary company, what our fees were and how much money we had saved him in rates for the ten years 1963 to 1973. Aubrey later rang me to say that we had got the job and said that I might be amused to know why:

'My Chairman, Mr Clore, looked at the bottom line of your schedule and when he saw the amount in fees that he had paid you tears ran down his face. Then he looked at how much you had saved him in rates over ten years and his tears turned to whoops of joy.'

'Give them everything we've got!' cried Clore. 'They're saving us a fortune.'

'So,' Aubrey said, 'I have been told to instruct you to deal with the rating of all our properties throughout the United Kingdom'.

Feeling a bit cocky, I said:

'That's nice, but why not America, then?'

I cannot leave the subject of Charles Clore without telling one story I heard about him – at a time when he was a household word throughout Britain and the undisputed king of the property world. He was dining at the Savoy Grill and some pushy, young, aspiring estate agent came up to him and asked if he could do him a favour. He explained that he was waiting to dine with someone who might be an important new client with whom he hoped to pull off a big deal and he wanted to impress him with his property connections. As Clore was leaving, would he come up to his table and greet him, his name being Jeremy?

Clore was always happy to oblige, so long as it did not cost him any money, so in due course on his way out he came up to the table where Jeremy was chatting with his guest and said: 'Hullo, Jeremy. Nice to see you. How are you?'

Jeremy said: 'F*** off, Clore. Can't you see I'm busy!'

I got married on 2 April 1955 and I suggested to my wife that if she wanted to see me a bit more she might like to accompany me round the shoe shops. She readily agreed and we had a very pleasant summer going round the south coast resorts referencing Mr Freeman,

The author and his wife, Sue, on their wedding day, 2nd April 1955

95

Mr Hardy and Mr Willis and the like. Sue would hold the clip-board while I went round the shop with my five-foot rod shout-ing out the measurements to her. When we had finished a shop, I'd ask Sue what she thought it was worth. I wrote down the rental figures and it later proved that she was a remarkably good valuer.

When the shops shut at 5.30 p.m. we would go to our hotel and, over a drink, work out the floor areas. Then we'd change for dinner after which we might go for a stroll along the prom-enade to sniff the seaside air and look at the waves breaking on the shore. It was a very pleasant summer, though by late autumn Sue was looking rather pregnant – if, that is, you can be *rather* pregnant. Perhaps not. I never charged the firm for Sue's professional services which, now I come to think of it, was pretty decent of me.

When I started negotiating shop assessments with the Valuation Officer I found it rather enjoyable. I like to think I have a fertile brain and have always regarded myself as an 'ideas man' rather than an administrator. I am also convinced that I would have made a much better advocate than a valuer. To obtain a reduc-tion in an assessment you had to think up some reason why it should be reduced – and a fertile brain is very useful for that. Once you had thought up the reasons, you had to convince the VO of their validity – and the power of advocacy is very useful for that.

The drill was this:

First, you had to convince the VO that your valuation, or some compromise figure, was either better than, or as good as, his. Therefore he could, with propriety, reduce his valuation.

Second, you had to make the VO *want* to reduce his valua-tion. The way to do this was to make him feel unreasonable, for the one thing that every civil servant, especially a VO, likes to think is that he is reasonable. I once found myself at lunch next to the Chief Valuer of the Inland Revenue Valuation Office. I mentioned this and he said emphatically: 'I don't want my Valuation Officers to be reasonable. I want them to be right.'

You could also make a VO want to reduce his valuation by making him feel that he wasn't being 'much fun'. I remember one hot summer's day, half way through negotiations, I suggested we went out and bought some ice creams. We did. He finished his large cornet well before I did and I said to him: 'Look, John. Here's my final offer: £5,250 and a lick of my ice cream.'

He found the offer irresistible. Another offer which often proved irresistible was to say: 'Look. I'll be generous to a fault, £5,250 and for that I'll throw in a picture of Baden-Powell.'

I used that expression quite often and to this day occasionally get postcards from the far corners of the earth which have on one side a picture of Baden-Powell. Why 'B-P'? Well, why not? After all, he was Chief Scout.

In the West Country once, the VO and I got a bit stuck negotiating, so I said:

'I'll tell you what I'll do. I'll toss you for it. £8,200 or £8,250. Call!'

'Hold on a jiff!' said the VO. He got up, opened the door and looked up and down the passage. He came back and closed the door firmly behind him.

'OK', he said, 'but I had to make sure the coast was clear. My chief's a Plymouth Brother and doesn't approve of gambling. But don't let's muck about. We'll toss for £8,000 or £8,500. Call.'

A ploy I did on more than one occasion was to get the VO to agree to toss for it. I would then put my hand in my pocket, feel which side of the coin was Heads (it's easy – Heads is smoother), take it out of my pocket, put it 'heads' down on the back of my other hand and say: 'Call'.

If people call quickly without thinking, they almost invariably call 'heads'. If you say to them: Heads or Tails?' people call Tails. So the VO calls Heads and I win the toss. I then explain the trick to him and suggest we do it again, only this time we'll actually toss the coin up in the air. For some unknown reason, the VO always declined my offer to toss properly and said that I could have my figure in any case. After this happened a couple of times, I decided to abandon the practice as, if not exactly immoral, it was certainly rather near the knuckle. Nonetheless, the psychological reason why this procedure always worked is interesting to ponder on.

The important thing about negotiating was not to create an issue. If you did, the VO, being only human, would be reluctant to climb down. The valuers I dreaded most were the ones who were thick – not the Valuation Officers themselves – they were a bright lot – but a few of the Inland Revenue valuers at the Valuation Office. If you had a good case and the valuer was too thick to realise it, that meant you had to resort to the Local Valuation Court which took an enormous amount of time and trouble, was stressful, and used up a lot of nervous energy. And the fee you charged never really covered the expense.

One partner, Billy Oliver, was negotiating the assessment of a property and the VO refused to admit that in law it was 'industrial'. If it were, then the occupier would get seventy-five per cent relief from rates. Billy's patience was exhausted. Furious at the VO's stupid stubbornness, he got up, threw his papers into his briefcase and slammed it shut. As he strode to the door he turned and said:

'Luckily, the House of Lords can decide,' and he went out slamming the door behind him. When the reverberations down the corridor had finally died down, the door opened very slowly. A now timid voice said: 'Er... I think I forgot my umbrella.'

Chapter eleven

'Lies, Damn Lies, and Statistics'

One day, one of the partners, Tom Dulake, said he had a book that he thought I might find interesting. He handed it to me; it was called *How to Lie with Statistics* by Darrell Huff (Victor Gollancz 1954). I read it and found it not only very amusing but most enlightening. I was not the only one, for by 1967 the book was on its Ninth Impression. The title is brilliant but the title to this chapter, a quote from the Victorian Prime Minister Benjamin Disraeli, is not. Statistics themselves do not lie but you can obtain and present statistics in such a way that people may draw the wrong conclusions; however, you can come very close to lying by quoting statistics and yourself stating the wrong conclusions.

Ever since I read the book I have been fascinated with statistics and used them to great advantage in my career, as you will see in due course.

I had only done Elementary Maths at school (I couldn't manage Higher Maths) and so had never heard of 'means, medians and modes'. I learnt that the 'mean' was found by dividing the total by the number of items, the 'median' was the figure in the middle, and the 'mode' was the number that occurred most frequently. So, if you take these numbers, 1, 1, 2, 3, 3, 4, 5, 6, 6, 6, 7, the mean is 4 (44 divided by 11), the median is 4 (five numbers are lower and five are higher), and the mode is 6 because there are more 6s than any other number.

There are no firm rules as to when you should use which of these three statistics, so people often use the one that helps them most to prove what they want to prove. It can be tricky if you use the wrong one. Consider this example:

A builder bought some land on which to build a housing estate. He found the average (mean) occupancy rate of houses was 2.8 persons per house – so he starting building two-bed-room houses. They didn't sell. Why? Because those people with no children or grown-up children only wanted one bedroom, those with two or more children wanted houses with three, four or five bedrooms, and those with one child wanted more than two bedrooms because they expected to have another child. The builder would have been better without an average, for nobody wanted a two-bedroomed home.

Take this statement: 'Half the people in England are above average intelligence and half the people in Ireland are below average intelligence. So the English are brighter than the Irish.'

Well, that half of the English may be brighter than that half of the Irish, but the other half are not as bright as the other half of the Irish; nor, now come to think of it, do they have the gift of the blarney.

Everyone should be taught statistics because they will find it invaluable in everyday life. If taught properly it is a fascinating subject for everyone. People are fooled or misled almost daily by the selection and presentation of statistics. The chief culprits, I think, are people selling investments. They show graphs to demonstrate how much their investment has grown over a certain period. During the course of a week you may find advertisements in newspapers showing how much investments have increased in value over the years. Now look and see how the period of years they use varies. Some show the last four years, some the last five and others six, ten or twenty. Why have they all chosen different periods? So as to present their figures in the most favourable light.

Take these figures for a company's profits in millions of pounds for the last ten years:

12, 10, 8, 4, 6, 8, 9, 14, 11, 12.

Here are a few facts:

Profits have gone down by a seventh (14%) over the last two years.

The company makes no more profit than it did nine years ago.

Profits are up by fifty per cent over the last seven years

Profits have trebled in the last six years.

Perhaps, says the Chief Executive, we should mention the last six years in our advertisement. And that, of course, may be after the 'profits' themselves have been massaged by what is known as 'creative accounting'.

My awakened interest in statistics was soon to come in handy. I had to deal with the rating assessment of two shops in Union Street, Aldershot (see plan) which was then the principal shopping street in the town. I went to see the VO to try to negotiate lower assessments, suggesting the shops contained various disabilities, but the VO (the Chief, not one of his valuers) was having none of it. He was a bluff Yorkshireman and was quite unwilling to give me a small reduction in the assessment so that I might 'go and play somewhere else'. I realised that my case was too weak to take to the Local Valuation Court so I asked him for the justification for the Zone A price he had applied to Union Street. He handed me a schedule showing all the rents

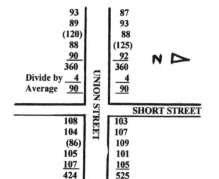

PLAN OF UNION STREET, ALDERSHOT
Shop rents for Zone A in new pence per square foot

Geographical analysis of Zone A Rents in Union Street, Aldershot for Zone A in the street. The average (mean) was shown at the end. It came to exactly twenty shillings (£1) for Zone A, which was the price he had used for assessing the whole of Union Street. Oh dear! I thought. I don't see how I can possibly get my shop assessments reduced.

Returning to the office next day, I thought I would start playing with the figures on the schedule. Although they were in shillings and pence and averaged twenty shillings, I will, to simplify matters, use new pence, a hundred new pence equalling twenty shillings. The rents for Zone A were somewhat on these lines:

93, 104, 87, 103, 125, 88, 108, 88, 107, 109, 90, 107, 105, 92, 120, 93, 89, 101, 86, 105.

The total was 2000 and there were 20 of them, so the mean was 100.

I then decided to separate the rents on the south side from those on the north. The south totalled 990, which averaged 99, and the north 1010 which averaged 101 – a difference of 2%. No joy there.

How about comparing the top (west) end of the street with the bottom (east) end? This could be interesting as my two shops were in the top half. Short Street intersected Union Street halfway up, so I separated the rents above the intersection from those below it. The result? They averaged 96.5 above Short Street and 103.5 below. Now we were beginning to get somewhere, but a 7% difference was not all that clear-cut. Perhaps I should stop calculating 'means' and look at the 'modes'. The top end rents were:

93, 89, 120, 88, 90, 87, 93, 88, 92, 125.

Eight of them were in the 87-93 bracket and the other two in the 120–125 bracket. If I ditched the two highest, the mean came down to 90, each side of the street also averaging 90. Now I was getting somewhere.

The rents on the bottom end were:

108, 104, 86, 105, 107, 103, 107, 109, 101, 105.

Nine of the ten rents were in the bracket 101-109 and the remaining one was only 86. Leaving out the 86 brings the mean up to 105.5 (106 on the south side and 105 on the north). I now have a 15.5% difference. Eureka!

The Local Valuation Court was in a few days' time so I got down straight away to preparing my case. These Courts are comprised of laymen, or laywomen, so I decided to present my case as simply and clearly as possible. I enlarged a street plan of Union Street and wrote the Zone A rents in front of each shop. The rents to be ditched I wrote in red. I then showed on the plan in large figures the average (without the ditched rents) for the shops above and for those below Short Street.

Apart from my street plan, which spoke for itself, I mentioned two points and two points only – 'keep it short; keep it simple' said one of my partners, Tom Dulake ('TSD')

My first point was to bolster up my reason for ditching the three rents. I pointed out that, under the relevant statute, the rent to be determined was not the actual rent but the rent at which the shop 'might reasonably be expected to let'. Another tip TSD had given me was that if I was going to quote from a statute or a court case, it was a very good idea to tell the Clerk of the Court beforehand so that he could have it ready. This I had done and as soon as I said I was going to refer to the statute, the Clerk handed the statute to the Chairman at the relevant page. This action of making the Chairman look at the quotation, and passing it to his two fellow members, obviously makes much more impression on the Court than if they just hear me quote it.

I suggested that if you had knowledge of the eight lowest rents on the top side ranging from 87 to 93, and was asked to guess the rents of the other two, you would not guess in a month of Sundays that the other two shops were let at 120 and

125. That was because they were not 'reasonably expected' rents. The same thing applied to the single low rent of 86 on the bottom side. For that reason, these three rents should be discarded.

The second point I made was that everyone knew that the favourite spot in a shopping street for a trader was next to Woolworths (who then had about a thousand branches). The difference in rents therefore between the top half of Union Street and the bottom half was precisely what one would 'reasonably expect'.

The VO then got up and said his piece. Then I cross-examined him: I asked him if he agreed that every shopkeeper's favourite position was next to Woolworths; he had to agree. Knowing this, why had it not occurred to him that he should compare the rents the Woolworths end with those the other end? He soon went red in the face and started blustering. I asked more questions, very politely and gently, and found the more polite and gentle I got the more the VO blustered. I confess that I really enjoyed it. After all, he was the big chief and I was just a young, inexperienced surveyor taking his very first case to the Local Valuation Court.

The Chairman, who I think was also Chairman of Surrey County Council, said they would now retire. They came back in a few minutes and said they would reduce the basis of assessment for all the shops in the top half of the street by ten per cent, i.e. to 90, which in actuality meant eighteen shillings per square foot for Zone A. I was thrilled. I had really enjoyed working out how to present my case in the simplest, most graphic, and most convincing way possible – and how to demolish one's opponent's case by cross-examination

For the rest of my career, I always enjoyed conducting a case far more than being a mere expert witness, but I could only conduct cases at the Local Valuation Court. A surveyor was not allowed to conduct a case in the Lands Tribunal (unless he was the ratepayer) and in practice did not do so at local public enquiries into planning appeals. In these, I was simply an expert witness giving evidence and then being cross-examined. I would have been a much better barrister-at-law than an expert witness. In other words, my talents were as a bowler rather than a batsman.

So started my career as a rating surveyor. For the rest of it, whenever I was negotiating, I used to imagine how my case and that of my opponent's would sound in court, and how they would stand up to cross-examination. This ability to conjure up in my mind vivid images of the court proceedings was an invaluable asset in deciding in my negotiations when to accept an offer – and when to refuse one and resort to the court.

I enjoyed advocacy so much and was so convinced of my ability that I came home one day and told my wife that I would like to switch from being a surveyor to becoming a barrister. We had three children by then and my wife, being practical, as women tend to be, asked how much money I would earn for the next few years if I changed jobs. I thought for a moment: 'Er…now I come to think of it, nothing; in fact I'd probably have to pay them for the first few years. Forget it. It was just an idle thought.'

Chapter twelve

Amenity Begins at Home

For my entire professional career at Gerald Eve & Co.'s London Office, I lived in the Walton and Weybridge area of Surrey. I could commute from door to door in exactly one hour, and there was wonderful countryside within a few minutes' drive where you could walk a dog for hours without seeing a building. I was truly 'suburban man', being fortunate enough to be able to buy a detached house 'standing in it own grounds' as the estate agents say. (In whose grounds do you expect it to be, for God's sake?). I was indeed the archetypal 'who's for tennis?' sort of chap, enjoying the grass courts and other amenities of St George's Hill Lawn Tennis and Croquet Club.

A friend of mine, John Guillaume, a local solicitor, asked me to join the Weybridge Residents Association (WRA) – which I did. It was not a political organisation like so many of the Ratepayer Associations used to be, but purely an amenity society. I found it most interesting – and a great help to me in the Town and Country Planning side of Gerald Eve & Co.'s practice.

One of the first things we did on my joining was to get out a Tree

John Guillaume when Master of the Solicitors Company

Book. We surveyed about a dozen public sites in Weybridge and decided which could be enhanced by the planting of a tree. The completed Tree Book was beautifully presented. On the left-hand page were an O.S. map of the site and a photograph of it as existing. On the other side was an indication of exactly where, how, and when the tree should be planted, its popular and Latin names, its characteristics, and its potential height and shape. Below was a photograph of the site on which had been painted by Brian Westwood, an architectural photographer, exactly what the tree would look like when mature. This magnificent book was then presented to the District Council who were so impressed that they ordered that all the suggested trees be planted. Indeed, they were so excited that they decided to plant trees substantially larger than those we had proposed, in order that the 'Westwood Effect' could be achieved more quickly.

Later, this book was given to Mr Duncan Sandys who had recently started the Civic Trust. I am told he stomped the country, armed with our Trees Book, urging other local amenity societies and local authorities to follow our example. It just shows how big an effect a few people can have on a country.

John Guillaume and I were later joined by Peter Preston, an architect, and the three of us prepared a document called *A Plan for Weybridge* which we had printed in large numbers and widely circulated. It pedestrianised the main shopping street and found almost universal approval by the membership. However, it did not find favour with the local planning authority who perhaps resented our trespassing on their territory and doing their work for them. To this day none of the shopping streets in Weybridge has been pedestrianised, with the result that their quality as a shopping centre has deteriorated and much of the trade has gone elsewhere. A pity.

Shortly after this, I became extremely interested in road design and the cause of road accidents. I read two fat volumes published by the Road Research Laboratory, which I believe is now called the Transport Research Laboratory. The two books were entitled *Research on Road Traffic* and *Research on Road Accidents*.

I learnt that you did not simply consider the *cause* of a road accident but the *contributory factors*; these could be both numerous and complex.

For example, a car comes out of a side road without stopping and is hit by another car. You might say that the cause of the accident was the driver of the car who entered the main road without looking properly before he did so. However, some of the following may have been contributory factors:

The driver from the side road:
did not see the halt sign as it was partially obscured by the branches of a tree,
was not familiar with this junction and thus unaware he had to stop – the white line and the word 'STOP', had been removed from the road surface due to resurfacing and had not been renewed.
The main road driver:
had his view in front limited by a hump-backed bridge,
was not paying attention to his front as he was looking at a pretty girl on the pavement,
was distracted by children arguing or fighting in the back,
was going too fast for the conditions,
had inefficient brakes,
had badly worn tyres,
should have been using his windscreen wipers or de-mister or de-froster,
had drunk alcohol,
had taken drugs,
was hurrying as he was late for an appointment,
was very tired at the end of a long journey
nodded off at the wheel for a second,
was not concentrating on the road as he was using a mobile phone,
The main road was:
wet,

icy,
snowy,
worn smooth.
It was:
misty,
foggy,
dark,
snowing,
raining.

Those are twenty-five factors which may have contributed to that accident. It is an accident that actually happened in Weybridge and the first three factors were in fact present in that case. Of course, there may have been more.

After reading these two books from the RRL, I realised that certain road improvements could reduce the number of accidents by a predictable amount – measured in, for instance, injury accidents per million vehicle miles. For example, 'improvement of street lighting' produced a reduction of thirty per cent of injury accidents in the dark. 'Improved alignment at bends' reduced injury accidents by 80%; 'provision of roundabouts' by 50%.

I was also very impressed by the beneficial effect if you converted a crossroads into a 'right/left stagger' whereby the driver on the minor road, in order to cross over, has to stop, turn right and then after a few yards turn left. This produced a saving of 60% in injury accidents.

Two years ago I wrote to my local Highway Authority pointing out a dangerous crossroads where there had been several minor accidents. I asked for it to be converted into a 'right/left stagger'. The Council declined to take any action and a few months later, there was a fatal accident there. Only then did the Council take action. I was appalled. It was so predictable.

The general public simply do not know that by failing to take action in certain cases, highway authorities are sentencing people to death. The only reason this is tolerated is that no one knows in advance precisely which people are going to die. The defence is often lack of money, but in fact many road improvements more than repay what is spent on them by the savings in medical, ambulance, fire, police and highway services alone. This is ignoring pain and suffering, inability to work, loss to financial dependants (which might qualify them for Government support) and damage to property. The trouble is that expenditure and savings come from different pockets. I have thought of a way of getting over this:

These days experts can assess with some accuracy the reduction in road accidents subsequent to road improvements and this can be expressed in financial terms. This can be set against the cost of the road improvements. Such a comparison is, I think, called 'cost/benefit analysis'. I suggest that if a highway authority can show the Government that a proposed road improvement is statistically likely to show a positive cost/benefit analysis, i.e. the savings during a certain period is likely to be more than the cost, then the Government should give the highway authority a grant for the improvement. Overall, it will save the taxpayer money.

Consequent upon my study of the contribution that road design can make to the prevention or road accidents, I got the Weybridge Residents Association to agree to make a local study of accident 'blackspots'. We found that the local Road Safety Officer, a retired policeman, on his own initiative made a blue blob on a local map where each accident had occurred in the last few years. It seems that no one bothered to study this map to see if the road design could be improved – they simply put it down to bad driving, bad road conditions or bad luck.

Off we went to study the blackspots – places where clusters of blue blobs showed there had been several accidents. I found it quite fascinating. Most of them were pretty obvious. I will give you a few examples:

A long, straight road with a fairly sharp bend at the end. There was no 'sharp bend' sign because the angle of the bend was not considered sufficient. However, in this context a sign was necessary because cars tended to travel fast and the bend was unexpected. Remedy: erect 'bend' warning sign.

A junction onto a main road (speed limit 60mph) where the main road had a hump just before it, obstructing the view; so if a car was going at 60mph it could not stop in time to avoid an emerging car. Remedy: remove hump.

A double junction where five roads merged. Incredibly, accidents here happened once a fortnight. Why? Because drivers were confused and found it difficult to look at four roads at once, before emerging. Remedy: Construct a roundabout.

We held a public meeting at a large hall, formerly the County Cinema, and it was absolutely packed. People had to stand at the back and along the side walls. We gave everyone a handout as they entered. This comprised:

Summarised statistics showing the typical savings in accidents that could be made by various alterations in road design.

A map showing the sites of the accident blackspots and how many accidents occurred at each.

A statement showing how in each case the road design contributed to the frequency of accidents.

Extracts from the O.S. map showing the site of each blackspot.

A description of the proposed remedy.

The Chairman of the Meeting, John Guillaume, then went through all this and showed photographs on a screen of the fifteen blackspot locations, explaining the problems and remedies for each. The Council's Engineer and Surveyor, who had been given notice in advance of the meeting, said he was 'on a hiding to nothing'. He said he was aware of the problems but it was difficult to get the Council to spend the money. One Councillor had said to him: 'There are no votes in roads.'

'Perhaps,' said John, 'the holding of this meeting might make it a little easier for you.'

It certainly did, for in a remarkably short time many of our proposed remedies were carried out. We haven't yet been thanked by all the people who would have been killed or injured were it not for our campaign – for the obvious reason that they do not know who they are. But they are out there all right, and I feel we achieved something really worthwhile.

Some of the things you read in the paper about research into road accidents are quite absurd. Only last week I read that speed was a factor in 50% of road accidents. Only 50%? Excuse me, but is not speed a factor in all accidents? Admittedly you sometimes hear of an accident where two cars collided and both drivers swear they were stationary at the time, but these must be quite rare, and one of the drivers might just possibly be lying or being a little 'economical with the truth'. But surely if there were no speed there would be no road accidents.

Ah, you may say; what they mean is 'excessive speed', but, regardless of statutory speed limits it could be held that any speed is excessive if there is an accident. And in any case who is to say what is excessive? At one time, any speed over walking pace was considered excessive and illegal (4mph, 2mph in towns) for a motor car ('horseless carriage'), because by law a man had to walk in front carrying a red flag. I understand this law was abolished in 1896 because so many of the wretched chaps walking in front with their flags were run over by the car behind. Perhaps these men in front stumbled into potholes or just collapsed with fatigue – and under the horseless carriage they went.

Even after this law was repealed and these unfortunate flag carriers were made redundant, any speed over 14 mph was thought excessive and was made illegal. Later the speed limit was raised to 20mph but this was abolished in 1920 when there was no speed limit at all until the 30mph speed limit for built-up areas was introduced in 1935. In the year 2003, the Transport Research Laboratory found that excessive speed made up just six per cent of definite contributory factors in accidents.

★★★

Shortly after this meeting I read in the local paper, the Surrey Herald, that Paddy, an Irishman would you believe it and one our WRA committee members, had been charged with driving under the influence of drink. It was alleged that he slurred his words. I immediately recalled that Paddy had a broad Irish accent and always slurred his words; this could easily be mistaken for drunkenness, so I contacted him and offered to give evidence to that effect. He accepted my offer.

In due course I attended the court and after waiting for hours was finally called into the witness box. I had given Paddy's counsel a short proof of evidence and he led me through it:

I told him that Paddy and I had both been members of the Weybridge Residents Association Committee for some years. Then counsel asked me:

'Mr Eve, how did you first hear of this case?'

'I read about it in the Surrey Herald.'

'And what, in the report of the charge, particularly caught your attention?'

'I read that they thought he was drunk because he slurred his words'

'Yes…'

'It occurred to me at once that Paddy always slurred his words and that this might quite naturally be mistaken for a sign that he was drunk.'

'Mr Eve, are you able to give an imitation to the Court of how the defendant normally talks?'

I thought: this had better be good; I had been practising it for ages. I took a deep breath, turned round to the magistrates on the Bench and put on a really slurred, thick Irish brogue:

'Hwell, your Honour, Puddy being Oirish, his voice shounds shomethin' loik this, to be sure, shir,' I said.

To my utter surprise the whole court, Bench, counsel, solicitors, jury, press and public, burst out laughing. I looked round in complete surprise. Had I said something funny?

Not only had I said something funny – I had got the man off.

***

Some years after the road accident meeting, I recall that at a WRA committee meeting one member said that we should try to increase the membership.

'What we need', another said, 'is a burning issue'.

'No problem', I said, 'It is easy to get a burning issue. All you need to do is mention a hospital. People are very touchy about hospitals.'

'Which hospital?' asked John, the chairman.

'Well, what about Weybridge Hospital?' I said. 'Perhaps it might close. There is some new plan in a White Paper recently published by the Government which proposes to close cottage hospitals and create great, big super-hospitals.'

'But we can't call a meeting about the closing of the hospital before we know it's going to close,' said another member.

I scratched my head for a moment. 'Yes, we can,' I said. 'We'll send a notice to our thousand members in the form of a telegram: *WEYBRIDGE HOSPITAL TO CLOSE OR NOT TO CLOSE? THAT IS THE QUESTION.*

Then we'll give details of the place and time of the meeting.'

The effect of this 'telegram' was beyond my wildest dreams. Not only was the meeting hall packed to capacity, including the permitted number of people standing, but scores were left waiting on the pavement outside.

John chaired the meeting in his usual capable and charming way, introduced Matron and a doctor and surgeon or two, each of whom said their piece followed by deafening applause. The meeting carried unanimously and with acclamation a strongly-worded resolution to be sent to the Government, opposing any possible closing of the hospital. All present resolved to do their utmost to campaign against its closure.

It must have put the fear of God into someone because, to this very day, in spite of big, modern hospitals in the district, Weybridge Hospital remains open and much loved.

Oh, and by the way, we increased the membership – by quite a few.

Q.E.D.

Chapter thirteen

A Barrel of Fun

My next visit to the Local Valuation Court concerned a barrel store in East London. In 1928, the Government, in an effort to boost depressed industry, passed an Act which gave industrial properties seventy-five per cent discount off their rates. To qualify as 'industrial' under the statute there had to be carried out on the property some industrial process, within which was included 'adapting for sale'. The VO concerned refused to agree that this building qualified as an 'industrial hereditament' and said that it was simply a place for storing barrels.

I went to visit the place and as I entered there were several men in the store rolling barrels around the place and a bit of hammering going on. I asked who was in charge and went up to chat with him. I explained I was trying to get the rating assessment down.

'What are those men doing with those barrels?' I asked.

'Them's not barrels, guv; them's casks', he said, 'and those geezers are coopers. They're skilled craftsmen; the trade is handed down from father to son.'

'Can't anyone be a cooper, then?' I asked

'Not bloody likely, mate! Not unless you're Gary Cooper, or your father was a cooper and his father before him, and so on. My family have been coopers since the time of Julius Caesar.'

'Gosh' I said 'Caesar was here' – I made a quick calculation – 'just over two thousand years ago. I suppose you and your family must have got the hang of it by now.'

'Bloody nearly, mate.'

'Why are these chaps here hammering?'

'Well, they are on tenterhooks, you might say.'

'Why is that? What are they worried about?'

'They're not worried about nothing, mate. These geezers have got a job for life. Look, guv'nor, do you want to know what exactly is going on here?'

'Yes, please; that would be marvellous.'

The foreman then explained to me that these casks were made of wooden staves held together around the circular piece of wood at the top and the bottom by metal hoops. The hoops were secured by tenterhooks, which are bent nails. When a liquid, such as beer or wine, is put in the cask, the wood slowly absorbs it and swells. The tenterhooks allow the bands to expand. When the cask is empty, the wood starts to dry out and contracts, leaving gaps between the staves. If liquid were poured into it, it would leak like a sieve. The coopers tighten up the bands round the casks using tenterhooks to make them watertight again, or rather, beertight or winetight.

'So when the barrels – sorry, I mean casks – come in here they are useless for anything? Yes?'

'Yes, guv.'

'And when they go out of here they are ready to be filled with liquid and then sold?'

'You got it, mate'.

Well, I thought, if that's not 'adapting for sale' in accordance with the 1928 Act, I'm a cooper.

I went back to the office delighted and, before I had time to forget anything, drew a rough sketch of a cask and wrote on it the names of all the various parts. I then started to prepare my exhibit for the Local Valuation Court – a careful drawing of the cask with all its parts named. I worked out I would need one copy for the Chairman of the Court, two for its members, one for the Clerk, one for the VO, one for me and one for the file – seven in all.

I read up the relevant Section of the statute – I think it was the Rating (Industrial Apportionment) Act 1928, but I wouldn't put money on it, and I'm certainly not going to go to the trouble of swotting it all up again; I've had enough of that. I think there was also a court case that defined 'adapting for sale', which held that it involved making something unsaleable into something saleable. That fitted my case perfectly, so I thought I had better take along to the court the textbook containing details of the case. I would also telephone the Clerk of the Court to tell him from what case I was going to quote so that he could have the relevant extract ready for the Chairman. Clerks really appreciated that.

In due course, I appeared at the Local Valuation Court and presented my case exactly as planned. It was great fun handing round my exhibit to everyone and then giving a little 'lecturette'. In essence, I said the building in question was not a 'barrel store' – it was a 'cooperage'. The court seemed most interested to learn about the coopers and their casks, staves, hoops and tenterhooks. To jazz it up a bit, I repeated my conversation with the foreman, and his witty remarks, particularly about Gary Cooper, got a laugh from the Court.

The court found in my favour. It was a finding of fact, rather than an opinion of value, so the rating assessment was automatically reduced by seventy-five per cent. It was only a small assessment but the clients were most impressed when I told them they would now only have to pay a quarter of the amount they paid in rates formerly. As you can imagine, as a young surveyor coming into his father's firm, I was anxious to prove myself; this victory was a great help to me in gaining some confidence in my ability.

When I got back to the office, I sat down and thought what I could learn from this success, for you can learn just as much from victories as you can from defeats. It's a bit like golf, really: Don't just think what you did wrong after a bad shot. Think what you did right after a good shot. So what did I learn?

1. Always try to get to know more about your subject than the opposition does.
2. Take great care to get your facts right. Check and re-check.
3. Make notes on the spot and don't throw them away – even when they have been typed out. The typist may have made a mistake. Later, I learnt how impressive it was, where there was a dispute as to fact, when my partner, TSD, said in the witness box under cross-examination: 'I would like to refer to my notes made at the time I inspected the property.' He asked for his file to be handed to him and he read out the relevant extract. Not only was he right but everyone was convinced he was right. Most impressive.
4. Try to simplify things. Everything is simple once all extraneous matters have been stripped away and the thing broken down to its constituent parts.
5. Give the court something to look at, not just something to hear. The more senses you can appeal to, the more your audience will listen, will understand, will remember. I learnt that in the Army.
6. Look at the members of the court to see what they are doing. Are they keeping up with you? Are they blowing their noses? Are they chatting to each other to clear up a point? Has a pen run out of ink or has it dropped on the floor? Are they pouring out a glass of water? All these things can reduce or eliminate concentration for varying periods and, if they have missed one point, then they may not follow the next ones you make that logically follow on from that.

I had really enjoyed conducting cases in the Local Valuation Court ('LVC') and thought I had better see how other more experienced people in the firm conducted them. So I went along with another assistant surveyor in the firm, Michael Hopper, to an LVC where he was appearing. He conducted his case in very clear, measured, confident tones and his cross-examination, though scrupulously fair, was very severe. I was impressed – and very glad I was not on the other side. Michael fulfilled his early promise by eventually becoming Senior Partner of the firm. When he retired from that he was appointed a member of the Lands Tribunal, a great honour for a chartered surveyor; the Tribunal's function was principally to hear appeals from the LVC and to hear disputes on compensation.

In a fit of enthusiasm and with a desire to better myself, I went out and bought a book called *The Art of Advocacy* by a barrister named Monkman. Unfortunately, I no longer have it, but I found it most instructive. However, looking back now, I think the two things that helped me most on my advocacy were first, translating Cicero at Stowe School, and second, a visit to the Old Bailey in London.

Marcus Tullius Cicero (106-43 BC) was the greatest of Roman orators and wrote treatises on rhetoric. By 70 BC, he was a prominent lawyer and seven years later was elected consul. He didn't rate much as a judge to my mind because he executed the Catiline conspirators without trial. I wasn't the only one either to think that that was bang out of order, because for that act of injustice he was exiled for eighteen months.

One of the passages we had to translate at school was called *In Catilinam*. Classicists will have noted that *In* is followed by a noun in the accusative, so that *In* doesn't mean 'In'; it means 'Against'; so it was Cicero's tirade against Catiline, which was probably more correctly called his opening speech for the prosecution.

I remember translating passages that said: 'he may say' this and 'he may say' that. Each passage was followed by the answer to that possible contention. The result was that by the end of his speech there were no new contentions to make that were relevant, and every contention that might be made had already been raised by Cicero, dealt with, and dismissed as nonsense. The effect of all this was that when opposing counsel stood up to make his speech there was nothing new for him to say. I thought this tactic was brilliant and I can recall one particular occasion when I put it to good use.

I was appearing before a LVC concerning the assessment of a retail shop. It had a new shopfront but was at the very (wrong) end of the parade. Position is all-important for a shop, the quality of the building being of secondary importance. You will make more money selling out of a suitcase in Oxford St, London, for five minutes than you would for five hours in a modern shop in a poor position. I was worried that as soon as I had sat down, the VO would get up and say what a wonderful new shopfront the shop had and that therefore the assessment was fair. So in my speech to the court I stressed the importance of position compared with a new shopfront. The three most important things about a shop were: 'location, location, location'. Just before I sat down I suggested to the court that when the VO got up he would talk about the new shopfront and that when he did so they should bear in mind what I said about position being of prime importance.

'Mark my words,' I said, finishing off, 'as soon as I sit down, the VO will get up and say what a nice new shopfront the shop has got. And, when he says that, you will bear in mind the saying handed down by Gerald Eve to all in this firm about valuing a shop: 'It's not what it is. It's where it is'.

This proved most effective and I got a reduced assessment. Just think how different it would have been if I had never mentioned the new shopfront. The VO could have got up and added an entirely new factor to be considered by the court. And the court would think: why hadn't Eve mentioned it? His valuation is all one-sided and should be rejected.

Now I come to my visit to the Old Bailey, properly known as the Central Criminal Court, which takes its nickname from the street it is in. The street itself takes its name from the Bailey, originally an outwork on the old City Wall, and the court itself is built on the site of the former Newgate Prison.

My brother Julian and I as youngsters went to the Old Bailey to see if we could find an interesting case to watch. A policeman told us that indeed there was one in a certain court which he was going to listen to during his break and suggested we went with him. We readily agreed. It concerned a mother who was charged with infanticide, which is defined in law as the killing of a newborn child.

When we came into the court the mother had just finished giving her evidence-in-chief and was about to be cross-examined. Up got a bewigged chap with big bushy eyebrows. He began:

'Now, Mrs Smith, I know it is very difficult for you, but I want you to do your best to try to remember exactly what happened on the evening of July 6th. I'll need to ask you a few questions, I'm afraid. All right?'

The witness sobbed a bit, wiped her eyes and blew her nose. She nodded. The bewigged man began, and a nicer, kinder man you could not imagine. Nonetheless, kindly though the tone was, the questions were highly material.

Our helpful policeman whispered: 'That man there is Sir Anthony Hawke, prosecuting on behalf of the Crown. The reason they are using him is that he can be very kind and gentle when it's appropriate. After all, if he bullied the poor mother, the jury would acquit her whether she did it or not, wouldn't they?'

Yes, they would, I thought, because once you've lost the sympathy of the jury, more likely than not they will find against you.

My mind wandered back to my LVC where I had been the aspiring, keen young lad taking on a blustering, red-faced, stubborn old Yorkshireman. I resolved never to come over to the court as a cocky, young clever dick or, as I got older and better, as a bully.

An instance of the usefulness of this resolution came to me near the end on my professional career. I was negotiating the assessment of an office building in Westminster and it became apparent that the assessment was far too high and should be more than halved. The Valuation Office valuer concerned, not one of the brighter nor more experienced ones, was not prepared to admit he had made such a big mistake, and he mentioned some office blocks which he thought comparable. He admitted to me that he had not seen them, even though the LVC hearing was in two days' time, because of extreme pressure of work.

A rather base thought went through my mind:

'How many appeals have you got outstanding?' I asked in a friendly, sympathetic way.

'Oh, about 5,000, I think.'

'You poor chap', I said and got up and left, the assessment remaining unagreed, but my *modus operandi* for the Court was now complete.

At the Court, but for Sir Anthony Hawke's example, my cross-examination could have gone like this: 'Mr Brown, you have quoted these other office blocks as being comparable but have you inspected them?

'No.'

'I find that quite amazing. I put it to you, Mr Brown that you have never even seen them!'

'Quite frankly, I haven't. I am up to my eyes in work; I have 5000 appeals to deal with and I simply haven't had the time. I would like to have had time to do so, but even though I've been working all times of the day and night, I simply haven't a spare moment. I can't call on lots of assistants like you can'

By this time, I would have lost the sympathy of the Court and been considered a big bully.

This is what Sir Anthony would have done and what I did:

'Mr Brown, how many rating appeals have you got outstanding at this moment?'

'About 5,000.'

'5,000! That's a lot. I suppose with all those appeals you must be incredibly busy?'

'I certainly am.'

'And being so very busy, it is perfectly understandable that, with the best will in the world, and I am not criticising you here for one moment, you haven't in fact been able to find time, have you, to inspect the other office blocks you put forward as being comparable with the appeal property.'

'That's right.'

'I am much obliged. No more questions, Sir.' I sat down.

'Thank you, Mr Brown.' said the Chairman, 'Would you now present your case, Mr Eve?'

'Sir, I am indeed lucky not to be in Mr Brown's unfortunate predicament in that through no fault of his own he has not seen the office blocks that he has quoted as being comparable. Fortunately, Sir, I have had time to inspect them all, and thus am able to help the court. I will now explain to you in detail why each of them is far superior to that of my clients...'

I won the appeal. What was I? Mr Nice Guy or a cunning advocate?

Chapter fourteen

A Day Out with 'JP'

A day out with JP was not to be regarded with any sort of equanimity, for the initials JP stood for no less than Captain Jack Powell, senior partner of Gerald Eve and Co., the ultimate head of one's working life, fearsome by reputation, fierce in manner, formidable in looks – and speech as clipped as his military moustache.

JP was, I am told, held in awe ever since he came to partner *the* Gerald Eve. The story goes that when JP first joined the firm, Gerald asked him to write a proof of evidence for him, as all the other partners did, and JP said:

'I only write my *own* proofs of evidence.'

The scene must have been reminiscent of one of those old Bateman cartoons, this one entitled *The Man who Answered Back Gerald Eve*.

My own claim to a Bateman Cartoon scenario was at JP's daughter's 21st birthday party at the Poole Harbour Yacht Club when, thanks to egging on by his son, Geoffrey, I became *The Man who Asked for a Port-and-Lemon at Poole Harbour Yacht Club'*. But I digress.

How come I was to have a day out with JP? Charles Clore asked Douglas Tovey, the shop agent who had suggested the Sears takeover bid, what to do about all the rating assessments which would be contained in the new 1956 Valuation Lists. Rumour had it that rateable values might be at least doubled or trebled, because the last rating revaluation had been in 1935. Tovey said

EC (Eric) Strathon (left), HWJ (Jack) Powell and his son, JG (Geoffrey)
Powell, at the RICS Annual Summer Meeting at Brighton, June 1953

to Clore, and he quoted it in his letter to JP: 'The best man in
England for the job is Captain Powell.'

So JP got the job and Douglas Tovey invited him to go up to
Leicester with him to meet the directors of Freeman, Hardy &
Willis at their Head Office. JP told me to accompany him; nice
of him I thought at the time, but much later I realized why –
because he was never himself to negotiate a single one of the
thousands of assessments that Clore eventually gave us to deal
with.

At 9.30 sharp, a huge claret-and-cream Rolls Royce drew up
outside the office at 6 Queen St Mayfair with, I think, a chauffeur similarly clad – in claret-and-cream. I joined JP and Tovey
on the enormous back seat and, nearly overcome by the delicious smell of leather, we sped silently and smoothly through
London to Leicester.

I remember Tovey introducing us to the directors of the shoe firm, having a desultory look at a bit of the factory and having an excellent lunch. All delightful – but to me it seemed a complete waste of time.

But The Day Out With JP, the title of this chapter, came much later.

I had by then referenced the Head Office, factories and warehouses of Freeman Hardy & Willis and Trueform in Leicester and Northampton, and JP told me that he was coming up to value them.

'You must get out and *value*' he used to keep saying and that is what he was going to do.

I collected JP at the office at 9.30 precisely and nervously drove him up to Leicester. We stood before the first building, a huge factory, and JP said brusquely in his usual clipped tones (he spoke rather like Field Marshal Montgomery):

'Right, Hilary, where's the referencing for this building?'

A shaky hand spends what seems like an age dabbling in my briefcase for the appropriate pieces of paper. At last I find them and, belatedly but proudly, insert them under the bulldog clip on my millboard. JP takes it, looks *down* at the first line, looks *up* at the vast roof and, with a 'tut', raising his head, his eyes looking up to the heavens, he tosses the millboard back to me.

I catch it and look down at it in dread. The entry reads 'Roof – pitched slate'. I look up and there is an endless expanse of – corrugated asbestos.

'Do it again and get it right next time' he says. 'It's no good my looking at your referencing any more.'

(Only later was I to learn that it had been re-roofed the previous week).

'We'll go inside now and *value*' says JP, turning on his heel and striding in to the front offices.

'All this at 3/6d [17½p]' he says.

I rustle amongst my pages of referencing to find the relevant page, at the same time hurrying along the corridor clutching my briefcase under my arm. I hope I have found the right lot of offices and scrawl '3/6d' in the penultimate column. By this time JP is halfway up the stairs.

'Half-a-crown,' he shouts from the first floor and races up to the second. '1/9d' a voice echoes down the stairwell.

I find it very difficult doing seven things at once: holding a briefcase, holding a millboard, holding a pencil, unclipping and clipping a bulldog clip, shuffling the pages of referencing, running along the corridor and, most important of all, writing the value per square foot on the right line in the right column on the right page.

For one brief, crazy moment I think of asking JP to hold my briefcase – but then I remember why he was first held in awe and think he might say: 'I only carry my *own* briefcase'. So I bustle up and down the stairs, along the passages, desperately trying to keep up. Ahead of me, JP seems to be going even faster. I can hear a stream of values echoing up and down the stairwells – '2/3, 1/6, 1/9, 3/6'.

I think: life is one long compromise. I can't carry on doing these six things at once. Something has got to go. I decide to write all the values on the top page – and sort out what went with what later. Although this sounded a good idea at the time, I later found that most of the figures were illegible as I had written them whilst running. It mattered not, as that night I completely failed to recall what value related to what.

Ask JP for his help? I felt pretty stupid right then – but I was not insane.

As you can imagine, we had soon got round all the floors of all the buildings and we headed off into the town centre for refreshment at some tearooms. We sat down at a table and discussed tomorrow's tour round the shoe factories at Northampton. The tea arrived. I was absolutely dying for some.

I poured out two cups and was just about to take a sip when JP said:

'Go and get the papers, Hilary'.

I remembered from JP's son, Geoffrey, that the Powells were avid newspaper readers and always took all three evening papers. I put down my cup and sped off. There was a newsvendor just outside and in the twinkling of an eye I returned, proudly proffering to JP The Star, the Evening News and the Evening Standard.

'Not those papers, Hilary! I mean the papers for tomorrow's visit to Northampton. We haven't done a day's work yet!'

I went off once again, crestfallen, returning with the Northampton referencing – and to some cold tea.

After tea, the newspapers left unread, JP asked me where we were staying the night.

'The White Horse, Leicester', I said.

Off we went, and on arriving at the car park at about 5 p.m. were asked by the attendant when we would be leaving in the morning.

'9 o'clock sharp,' JP told him.

This was rather later than commercial travellers started, so we were put at the very back of the walled car park.

At reception I said I had booked two single rooms under the name of Eve, but the receptionist said they had no record of it. JP asked if I had the letter of confirmation. I said I had and started looking for it in my bulging briefcase. No sign of it. I could hear JP behind me, puffing and blowing with impatient anger.

'It must be in my suitcase', I said.

I opened the case. The only paper on the top was a magazine, 'Playboy'. This produced a few nudges and low chuckles from some commercial travellers who had arrived by then. I unpacked my pyjamas, underpants, dressing gown, shaving kit etc onto the reception desk. As the pile on the desk grew higher and higher with the contents of my suitcase, the queue of people behind me grew longer and longer.

At last, eureka! There, at the very bottom of the case, was the confirmation letter. Even Mafeking was not so relieved as I was.

'There you are, two single rooms for tonight' I said, triumphantly, if somewhat belatedly, handing the letter to the receptionist.

He took it and looked at it.

'Yes indeed, Sir', he said 'but this letter is addressed to The Ram Hotel, Northampton.'

Without a word, JP, purple with rage, turned on his heel, barged his way through a long queue of laughing commercial travellers and swept out. With one broad sweep of the arm, I swept my belongings on the desk into my suitcase (taking, I was later to find, a couple of dozen brochures besides). I sat down heavily on the case, fastened it and ran the length of the hall, through the revolving door, out into the street and round the back to the car park. It had started to pour with rain and a wet JP, no longer purple in the face, but now white with fury, threw his arms sideways in a despairing gesture, presenting the car park to me.

'Now look what you've done!

It was not a welcoming sight. The walled yard was packed solid with cars, all carefully placed according to the stated time of morning departure. Mine, of course, was at the very back. Four cars had to be moved before we could get out. The time it took to record the car registration numbers, go to reception, discover the names of the car owners and their room numbers, trace all the owners (some of whom were taking a bath before dinner) and get them to move their cars in the correct order was the longest 45 minutes of my life.

My 'Day Out With JP' had indeed been the longest day.

When I got back to the office Bill Haddock, a salaried partner, helped me decide which of JP's prices might go with which floor space. All the valuations came to more than the rating assessments, so on Bill's advice I knocked off about 20% and went to see Bob Eeles in Leicester who was the contract valuer for the Inland Revenue.

Mr Eeles was an elderly gentleman and received me graciously. He said he had known Gerald well. After the pleasantries he suggested we got down to business:

'What's wrong with the assessments?' he started off, rather aggressively.

I explained politely that the buildings were old (I didn't mentioned the brand new asbestos roof), badly planned due to being developed piecemeal, and inefficient.

'All right then, Hilary, I'll give you ten per cent off the lot!' He got up from his desk and said: 'Now come and have some lunch'.

With alacrity, I accepted both offers.

There were two other memorable occasions with JP, and that of course is why I can recall them:

JP called me into his office and said in his usual brusque, business-like manner:

'Come in, Hilary. There's been a fire at the Ranelagh Club. Go and assess the damage. Here's the file.'

Off I went to the Ranelagh in south-west London which used to be a posh club before the war where polo was played. I think it was requisitioned during the war and had been handed back.

When I arrived I found a number of disused and derelict buildings. Vandals had obviously been hard at work. At least a dozen small fires had been started over the years. I could find none that was still smoking or had warm timbers. Where was the most recent fire? I had no way of telling. I went back to the office and told JP.

'Right, Hilary; can't do the job. I'll get someone else.' I was dismissed and I left the room with a burning sense of injustice.

Another time, I was called into his office. He sat at the far end of a long refectory table and in a fit of annoyance about some job I had done, he threw the file the length of the table and it landed in front of me. I heard a voice say:

'I'd like to know what I have done wrong.'

To my utter amazement I realised that it was I who had said that. He didn't answer but instead turned to talk to someone who had just entered the room. Clearly I had been dismissed. I went back to my office and burst into tears at the sheer injustice of it all. In a few minutes the intercom buzzer went. I knew it was from JP because his was the only buzzer that ran continuously until you answered it.

'Can you come up, please,' he said.

I put the receiver down without a word and got to my feet. As I climbed the stairs to JP's office, I thought that if he went at me any more I would tell him to get stuffed, Senior Partner or not. I entered the room ready to have a real go at him. He said: 'Hilary, I am going to throw something else at you now. Come and have lunch with me at the Bon Viveur.'

The Bon Viveur was a superb restaurant in Shepherd's Market and his invitation was his way of saying 'sorry'.

To be honest, he was not my sort of surveyor, but he ran the firm well and was very shrewd at getting the clients and keeping them happy. This was because he always saw things from the client's point of view. There were plenty of people in Gerald Eve & Co. who were so fascinated by the job itself that they forgot to keep the client informed and to write the sort of letter that would simply give the client the information he wanted. Instead they would give what they considered was a fascinating discourse on the technical complexities of the matter.

On more than one occasion, I saw JP read a long letter, written by some brilliant assistant to a client proposing to buy a property, and then write an accompanying letter on the lines of:

'Dear Freddie. Don't touch it with a barge pole! Enclosed is a letter giving all the technical guff. Yours sincerely, Jack.'

When JP wanted to give me some advice, he would preface it by saying:

'Tip from your father, Hilary.'

It would be:

'Tip from your father, Hilary: Always fold plans outwards.

'Tip from your father, Hilary: Never throw away your original notes, even and especially if they have been retyped.'

'Tip from your father, Hilary. Before you write on a bit of paper, always put the date and your initials in the top right hand corner'.

The tips were always excellent advice but whether they actually came from my father or not, I was never quite sure. I like to think so.

One other lasting memory I have of JP was a remark he made to me at a professional dinner when we had both gone to relieve ourselves during that period which conveniently is set aside for that very purpose just before the speeches. As JP washed his hands in the basin next to mine, he said:

'Hilary, the big difference between the French and the English is that the French wash their hands before having a pee, and the English wash them afterwards.'

I didn't give this remark much thought at the time, but over the years I think that JP's remark showed considerable insight into the difference between the two nations. The Frenchman thinks:

'I am about to handle the most important part of my body, so I will wash my hands before doing so.'

But the Englishman thinks:

'I have just touched something rather nasty so now I need to wash my hands.'

Chapter fifteen

'I'm a Simple Man'

'THOMAS SOWERBY DULAKE will prove…' was how TSD's proofs of evidence began. A number of assistants, including me, wrote proofs of evidence for him, one of the firm's partners, for he was much in demand; and it was a real joy to write them, too, because when you went to the Lands Tribunal, or a Public Local Inquiry arising from a planning appeal, to hear him giving evidence, he was superb. He was a very charming, gentle, courteous man with a fine intellect, and when he gave evidence in the witness box you felt he made the most of all the work you had put into his proof. At his peak, I have no doubt he was more in demand than any other expert witness on valuation and planning matters in the country.

The heading of this chapter is something that TSD said quite often. What it meant really is that he had such a good brain that he could reduce any complicated problem to its simplest terms, so that anyone could understand it.

When he asked you to write a proof of evidence, he would give you a piece of paper and on it he would write a very few words. When you had, after weeks of hard work, completed the proof, you'd think that you had done it all yourself, but you hadn't; for those few words on that piece of paper had chan- nelled all your thinking and efforts in the right direction. That was his method; that was his skill; that was his art.

On one occasion, when he asked me to write a proof for him, concerning a hearing by the Lands Tribunal of a rating appeal by Marks & Spencer, he wrote:

'4Z + Q.'

That indicated to me that he wanted the store to be valued on the zoning method using four zones and that at the end one should make a quantity allowance for the fact that the store was much larger than the other shops. On another Marks & Spencer appeal he wrote:

'Best evidence: rent of appeal property.

Next best evidence: rents of comparable properties.

Best evidence after that: rating assessments of comparable properties.'

For anyone conducting a rating appeal today, every word of that pithy summary remains true.

I remember TSD giving evidence at a Local Planning Inquiry concerning a proposed extension of a large wastepaper store in the Green Belt – an appeal which I thought had no prospect of success – and I told TSD so. Halfway through a very shrewd and severe cross-examination when TSD felt he was being manoeuvred into a corner by a brilliant barrister, Eric Blain QC, TSD swivelled round in his chair, turning away from counsel so that he faced the Planning Inspector, and, with a charming smile said:

'As I see it, Sir, it's like this…'

Counsel was powerless to intervene. TSD then explained the problem in a confidential tone, in its simplest terms, and in the most engaging and convincing manner.

'I am much obliged, Mr Dulake,' said the Inspector, eating out of his hand.

TSD turned to face counsel again.

'And what question was that answering, Mr Dulake?' asked the QC.

TSD said not a word, but, as he looked up at counsel, a broad grin spread across his face.

He won the appeal.

Another incident I recall was when TSD was being cross-examined by counsel in the Lands Tribunal. Counsel pointed out that there was a mathematical error in a calculation which had been submitted by us to the Tribunal well before the hearing. It was my fault and I felt really bad about letting TSD down.

If I had been in the witness box and suddenly been confronted with this error, I would have been really flummoxed, but TSD didn't bat an eyelid.

'When did you first discover this error?' he asked of counsel. Counsel wasn't expecting this question and turned to his surveyor. A whispered conversation ensued. This was pretty good, I thought: the witness cross-examining counsel.

'About a fortnight ago,' said counsel.

'Oh, what a pity you didn't tell me earlier! I could have rectified it, had it retyped and sent to the Tribunal well before the hearing so as not to waste everybody's time here today.'

Yes, we all thought. Why didn't they? What a nice chap TSD is – and what a nasty, unhelpful bunch the other lot are. That was another case TSD won.

My favourite story about TSD, one that he told himself, concerned a police speed trap. As you will shortly gather, it must have occurred quite a while ago. As TSD was driving along in his car, a policeman stepped out of the bushes at the side of the road and waved him down. TSD wound down the window.

'A very good morning to you, sergeant ' he said with his famous, charming smile.

'Good morning, sir,' replied the sergeant, pleased that he was addressed by his correct rank. TSD never missed a trick. 'This is a speed trap and I have reason to believe you have been exceeding the statutory speed limit of thirty miles per hour. I must ask you to wait while my colleague down the road comes to join me.'

TSD looked down the road; sure enough a police constable was bicycling along towards them. On arrival, he got off his bicycle and handed over his stopwatch to his sergeant.

'Thanks Fred,' said the sergeant. He looked down at the watch, showed it to TSD and read out the number of seconds.

'Now, sir,' he said, producing a clipboard, 'if you look at this table and at the column headed by that number of seconds, you will see it indicates that you were doing a speed of 36 mph over the measured distance of 150 yards.'

TSD thought for a moment.

'It certainly does, sergeant, but would you be kind enough to tell me how you measured the distance?'

'Fred paced it out, sir. It is 150 paces.'

'Ah, Fred paced it out, did he', said TSD. 'Well, Sergeant, the length of people's paces varies quite a lot, you know, mostly due to variations in height. For instance, I have a pace of exactly thirty inches. Fred looks to be of a similar height to me so probably he has a similar length of pace.'

TSD now used his quick brain to think at a speed well in excess of most people's speed limit. He continued:

'So if Fred's pace is thirty inches, not thirty-six inches, that means I would be doing a speed of about one-sixth less. So instead of doing 36 mph I would be doing 30 mph which is exactly the speed limit.'

Sergeant and constable looked at each other in utter amazement. There was a long pause.

'What do you do for a living?' said the sergeant.

'I am a chartered surveyor,' said TSD.

'Well, I've never met a chartered surveyor before,' said the sergeant and, as he turned on his heel to walk away, he looked over his shoulder and added:

'Bugger off!'

TSD's other favourite story concerned an old lady who lived in a village and received a bill from her local solicitor which read as follows:

£ S D

To seeing you on the other side of the street last Saturday morning.
To crossing the road to talk to you.
To finding out it was not your goodself after all.
To returning to the side of the road I was originally on   7s 6d

Chapter sixteen

'A Chapter of Accidents'
—Earl of Chesterfield (1694–1773)

Accidents will occur in the best regulated families... they may
be expected with confidence and borne with philosophy.

(Mr Micawber) *David Copperfield*

6, Queen Street Mayfair may have been roomy enough for the
mistress of the Prince Regent and her retinue but it was always
pretty crowded when I worked there – but I loved it. Everyone
was very friendly – I suppose it was so crowded you had to be
– and we helped each other with our problems. At one time
I worked in the basement and made photographic exhibits to
accompany proofs of evidence. Geoffrey Powell also did so and
we learnt about exposures, developing, fixing, printing and
trimming.

Once TSD gave us a vast aerial photograph which had cost a
fortune to take, £120 we were told – a lot of money in the early
1950s. He asked us to mark all the existing uses on the map. An
Existing Use Map was prepared by using standardised colours:
brown for Residential, purple for Industrial, green for Open
Space, blue for Land Covered by Water, etc. It was very nerve-
racking to do because we had to use coloured inks; so if you put
the wrong colour on a building it could not be deleted. When at
last we had finished it and become a couple of nervous wrecks,
we took it up proudly to present to TSD.

He looked at it. 'I am sorry, but I am afraid it won't do. Our
client wants to build on an open space and I want to present a

plan to the Planning Inspector showing there is plenty of open space around. Get another photograph and this time those areas you have coloured blue I want coloured green.'

'But,' I interrupted, 'the blue is Land Covered by Water and therefore should be coloured blue.'

'Hilary, I don't care what the land is covered with; colour it green, please' said TSD. 'After all, it's open space isn't it?'

Halfway through the morning we would go out and have a cup of tea or coffee at Ruggeri's in Shepherd's Market – a colourful place. On the way there we were invariably accosted in affectionate terms by a few smartly dressed ladies who, seeing we were busy men going for a quick break, thoughtfully asked us if we would 'like a short time, dearie'. I was intrigued by one of them as she used to wear an anklet which I was told meant she catered for women. One window in the market had a little red neon sign saying 'Private Tuition in French', but I was told it was not actually a language school, though nonetheless one might learn a thing or two about a foreign tongue if one went there. On one little noticeboard in the square was a notice: 'Polish gentleman gives all-over body massage – ladies a speciality'.

Ruggeri's was a wonderfully good but cheap restaurant run by an Italian family. When you went and sat at a long table you would find several of your colleagues there all talking 'shop'. As a result, everyone knew what was going on in the rest of the firm and helped each other with their problems. Much of the practice of Gerald Eve & Co. was dealing with problems that no one else felt they could cope with, and often provincial firms of surveyors and estate agents would refer a valuation to us because they had not the expertise to do the job themselves. For instance, how do you value a cattle market or an agricultural showground?

Estate agents were particularly fond of giving us that sort of job, because we did not practice estate agency in those days and would not therefore try to steal their clients from them.

Over coffee, someone might say:

'I've been given an oyster farm to value. Anyone done one?'

'Yes,' Billy Oliver might say. 'We did one before the War at Whitstable. We've still got the file. I'll look it up for you when I get back to the office.'

Bill Haddock said that in Gerald's day when there was something really tricky to value, like the value of some land required to extend the runway at London Airport, he'd gather everyone round the table and say:

'Now, gentleman, let us all air our ignorance.'

'It was easy for Gerald to say that,' said Bill, 'because after nearly fifty years of varied practice he knew damn near everything.'

The idea of the remark however was to encourage people not to be shy of coming forward with suggestions even if they were impracticable. It was what was later called a 'brainstorming' session – something which I learnt to use to my great advantage – particularly in the first leasehold reform case to come before the Lands Tribunal –but that is a story I will save for later on.

Not all the assistants at the office worked hard all the time. One chap called Hubert caused havoc to the typing of letters by one poor secretary: he magnetised one of the moving parts of the typewriter which had a metal letter on the end. Not content with that disruption, he got hold of another secretary's typewriter so that instead of typing 'r' it typed 'q' – with the result that her letter started something like this:

'Deaq Siq,
Qefeqqing to youq letteq qe Huqst Paqk Qacecouqse...'

It took the poor girl quite a while to realise that she had not suddenly lost the art of touch-typing.

Another incident which soon became famous occurred one evening when one of the partners went up to the top (fourth) floor to see if a proof of evidence required that night was nearly finished. It was after 'office hours' and it was not usual for a partner to venture to so high a floor. He was just checking the proof when suddenly the lights went out. The voice of the partner's young assistant surveyor, John Dunne, said:

'OK, girls, seduction time!'

One of the worst things that ever happened in the firm when I was there occurred when we were doing a vast valuation of some firm's assets. In those early days we had a calculating machine called a Monroe which was mechanical and had a large number of levers. We used this machine for adding, subtracting, multiplying and dividing and had a very high regard for it.

By working late for some days we just managed to meet the client's deadline: the Company Secretary wanted the total in time for the Chairman to announce it at his Board Meeting. One of the partners rang the Secretary and told him the total asset value. Overworked and underslept, we all fell back in our chairs exhausted. Then one chap idly tapped a sum on the Monroe, like 2×2. The answer that came out was 5.

'That's funny', he said. 'The Monroe got that sum wrong.'

'It couldn't have done' said the partner; I've used it for ages and it never makes a mistake. You must have pressed the wrong key. Let me have a go. He did another sum and the wrong answer came up.

'Oh, my God!' he said, 'And we've just given the total value to the Secretary, and he's given it by now to the Chairman and he's now given it to all the Directors. And it has probably been released to the press. And it's wrong. What the hell shall we do? For a start, we'd better add up all the figures ourselves and see how big the error is.'

This we did and, although some of the Monroe sub-totals were way out, we found that, by the grace of God, the errors in the long run all but cancelled themselves out. A few very minor adjustment were made to one or two valuations which we felt were warranted – valuation is not an exact science; in fact it's not a science at all; it's an art – and the total was precisely what we had given to the client.

Phew!

Each department had its own way of coping when mistakes were made. – not that they occurred often. Partners and Associates signed their own letters personally, but any letter, even a mere acknowledgment, written by an assistant was first initialled by him on the copy and then signed by a partner. No doubt that is why during the whole of my time at Gerald Eve & Co. we never had a negligence claim made against us. We used to get cross when the insurance premium against negligence claims went up every year, thanks to the carelessness or incompetence of other firms. We felt we should have got a 'no claims bonus'.

In my department, I noticed that the drawer of one filing cabinet was labelled 'OFFICE LUGER', presumably for use by those who had made a mistake and wished to end it all. However, when I opened the drawer, there were only files in it.

My own defence against my carelessness or stupidity was to start a Blundering Fools Club (BFC). When someone in my department erred in that way I suggested they joined the BFC. When I told them I was a founder member, it made them feel better.

I remember on one occasion I made what I thought was an awful mistake. I forget what it was – I am better at recalling my successes than my mistakes. I went into Geoffrey Powell's office to share my trouble with him – 'a trouble shared is a trouble halved'.

'What on earth shall I do?' I said. 'Sears is the most important client I've got.'

'Ring up Aubrey and tell him exactly what you've done,' said Geoff.

'Really?'

'Yes, really – and do it right now.'

I decided to take his advice, went back to my office, took a deep breath, lifted the receiver and dialled Aubrey Hawkins, the Property Director of Sears. I explained to him quite bluntly the stupid thing I had done and apologised profusely.

' Good Lord, don't bother to apologise, old boy. I made a much worse mistake than that only last week. Don't give it another thought but, Hilary, I do appreciate your coming clean and telling me about it straight away. The thing I can't stand is people making a bloomer and then trying to bullshit their way through it. It means they're taking me for a fool, which I like to think I'm not – most of the time.'

So there's a bit of good advice for anyone less than perfect in this world. After all, as Alexander Pope once said: 'To err is human, to forgive, divine.'

With that little proverb in mind, I will tell you the tale of The Old Lady of Threadneedle Street, the name for the Bank of England in the City of London.

The Bank of England instructed Gerald Eve and his partner, Jack Powell, on some matter concerning their premises in the early 1940s during the war. They visited the Bank and said they would need the plans in order that they could scale them off.

'I am sorry, Sir,' said the client, 'but for obvious reasons we never let the plans of the Bank of England be taken out of the premises.'

Now Gerald was a very formidable man when roused and he indicated in no uncertain terms that he was someone who could be trusted with valuable items. He said that if the official were unable to authorise their taking the plans then perhaps he would kindly go and get someone who could.

Off went the frightened little man and came back with his superior. After some discussion, the superior caved in – he was not the first to give way under the articulate, verbal onslaught delivered by Gerald in stern, measured tones.

'Just this once, then, Mr Eve' he conceded.

The two of them left with the plans and took a taxi back to the office.

Gerald ordered two cups of tea and sat down at the table with JP.

'Right, Jack, get out the plans and we can start scaling them off straight away.'

'But, Gerald, I though *you* had the plans.'

'No, I haven't got them.'

They both froze for a moment and then clapped their hands to their foreheads as the sheer enormity of what they had done sunk in. They had left the plans of the Bank of England in the taxi.

There is a lost property office for things left in London taxis, but drivers tend not to deliver to the office straight away items left in the back of their cab. Understandably they wait till they have got a job that takes them near the office. So when JP rang the office there was no sign of the plans. The trouble was that Gerald and JP had arranged to inspect the bank in detail the very next day. To inspect the whole building without plans was possible if one adopted the usual procedure of starting on the left at the very top. The trouble would come when they inspected the vaults, which were extensive and could fairly be described as a rabbit warren.

Gerald told JP that the only way they could get round the vaults without becoming hopelessly lost was for him to concentrate on nothing but navigation. He must surreptitiously make a note on his clipboard of every twist and turn they make, talking to no one for fear he might lose his concentration. Even a momentary lapse could be fatal to their chances of ever finding their way out again.

When they arrived at the bank, the official expressed surprise that they had not brought the plans with them. Gerald explained somewhat irritably that as experienced surveyors they were perfectly able to find their way round without being cluttered up by a lot of plans. My sister tells me that Gerald told her that wherever he went in the Bank a man followed him. This irritated him no end and he told the man that as a surveyor of some fifty years experience he had never been followed around a building before. The man said he was sorry but that was their invariable practice and not even Gerald could dissuade him from doing so.

JP deserves great credit for successfully navigating the three of them round the bank. Vasco da Gama could not have done better. Without JP's devotion to duty all their skeletons might even now be lurking in some distant vault between rows of gold ingots.

The two of them had to wait three long days before the plans were handed in to the lost property office. In due course, the plans were scaled off and returned, the Bank being none the wiser. Indeed, until or unless they read these memoirs, they will not yet know of their loss.

No doubt Gerald and JP were frightened that during the three days that the plans were missing they might have fallen into the hands of some bank robber who promptly made copies of them. If so, they or their descendants who are planning to make a raid have certainly been biding their time. Perhaps the robber has bequeathed the plans in his will:

> *And lastly, I leave the plans of the Bank of England to my son,*
> *Fred, in the hope that he can put them to better use than I did.*

Chapter seventeen

Butterflies and Hangovers

Shortly after the 1956 Rating Revaluation, the Multiple Shops Federation held a meeting of rating surveyors who acted for, or were employed by, multiple shop companies. TSD and I attended the first such meeting. There we all decided in which towns we thought the level of assessment in the multiple trading positions seemed excessive. A firm of surveyors was appointed to look into each one and Gerald Eve & Co. was given, amongst others, Coventry. Accordingly, TSD and I paid a visit to the city to look into the matter.

The German Luftwaffe heavily bombed the City of Coventry on 15 November 1940. The cathedral and much of the city were destroyed and 568 people were killed. It is said that Churchill, having broken the German code, knew that Coventry was going to be raided that night, but that if he had taken full advantage of such information the Germans would have realised the code had been broken and would have changed it. Whether that is true or not, it seems to be a perfectly valid reason for Churchill acting as he did. In any case, in those days the Royal Air Force and the Army anti-aircraft guns could do precious little to shoot down enemy aircraft on night raids even if they knew they were coming.

On our inspection we found that the principal shopping centre had been rebuilt. It was approached appropriately by passing an equestrian statue of Lady Godiva, naked but decorous – thanks to a head of long hair judiciously placed by her sculptor.

The Zone A price for the most valuable parade of shops was 50/– (£2.50), which we decided, was not justified by the rents passing under the leases.

TSD decided to consult Michael Rowe, QC. The two had worked together countless times and had become close friends. Indeed, they were so close, that on one occasion before a hearing, Michael said: 'I'm going to tear up your proof of evidence and do it another way, Tom. Don't worry; I'll lead you'.

Sir Michael Rowe CBE QC,
President of the Lands Tribunal
1966–73

You are, of course not allowed to lead your own witness – i.e. intimate to him or her what answer to give, but such was the close professional relationship between the two of them that Rowe was able to do this without anyone being aware of it.

Off went TSD and I to my first 'con' with counsel, at Rowe's chambers, 2 Mitre Court Buildings in the Temple. We called it a 'con' because if you were going to see a QC it was known as a 'consultation', but if you were going to see a mere 'junior' it was a 'conference'. So 'con' did for both. We arrived a bit early because TSD and I had, at his insistence, set off in his chauffeured car rather early.

'Never be late for a con with counsel, Hilary' said TSD. I took his advice and although I was to have scores of cons during my career, I was never late.

Counsel's chambers were not a bit like I imagined. I thought that with the amount of money top QCs made they would house themselves in luxurious offices. Not a bit of it. An open, stone-faced entranceway contained just inside it, painted in black letters on the wall, a list of all the barristers on each floor. A very worn, stone staircase with metal handrail and balusters led you to the upper floors. There was no lift. As you entered the suite of chambers you went into the clerk's office which was very cramped and absolutely chock-a-block with briefs, each tied up with pink tape. Sometimes you had to stand waiting for the others to arrive but if you were lucky you might be able to sit in one of the barrister's rooms which was at present not in use.

The clerk was a very powerful cog in the chambers' wheels for it might well be he who decided which briefs to accept and how much to charge. It was also often he who would decide which of the young barristers to employ on a particular case, either on their own or as a junior to a QC, otherwise known as a 'silk'. In those days, if you employed a silk, not only did you have to pay his fees, but also the junior, whom he had to appear with, who was paid a fee equal to two-thirds that of the silk. Barristers had therefore to think twice before applying to the Lord Chancellor to 'take silk', as they had to guess whether their income would increase or decrease by their so doing. Such a step was termed 'taking silk' because on appointment the barrister 'exchanged his stuff gown for a silk one'. No barrister was permitted to be instructed direct, so you had to pay a solicitor's fee as well, for briefing counsel and for attending the proceedings in the Court, Tribunal or Planning Inquiry.

The solicitor was often completely redundant, for when Gerald Eve & Co. submitted a proof of evidence to the solicitor, we knew far more about, for instance, the law of rating than he did and had usually done all his work for him; all he had to do would be to put a 'backsheet' on our proof, stating 'Counsel to advise', tie it round with a bit of pink tape, deliver it to counsel – and charge the client one thousand guineas. This amount was often more than we were charging for months of highly specialised, skilled work.

Coming back to our case, TSD saw the clerk and we were shown into the room of Michael Rowe who, incidentally, later became President of the Lands Tribunal and received a knighthood. He was a very able and likeable man, most approachable, with no side to him whatsoever.

After discussion, TSD and Michael Rowe decided that neither of them would appear at the Local Valuation Court hearing the Coventry appeals. They would 'put the boy in'. I was the boy.

I prepared my case and went up to Coventry. This being, I think, the first LVC hearing of a test case concerning a new, post-war shopping centre, a large number of rating surveyors attended together with several solicitors. Many years later, I remember, before giving evidence in some important case, saying to John White, the Surveyor of Marks and Spencers, that in my younger days before I had a reputation to keep up, I was never nervous.

'Oh yes you were, Hilary,' he replied 'I remember many years ago standing next to you at the gents urinal when you were just about to appear before the Court in the test case at Coventry and you said to me "And to think, John, I used to *collect* butterflies".'

I opened my case, gave my evidence and was asked a few questions by the VO. He then gave evidence and my cross-examination must have gone well, because during it I heard the solicitor, Bill Everett of Lovell, White and King, whisper to his next door neighbour:

'This chap's in the wrong business.' I rather think he was right.

In due course, the court retired to make their decision. The general feeling amongst the solicitors and surveyors was that the day had gone well and that the court would give a reduction of 2/6d (12½p) or 3/6d (17½p). We were therefore all very surprised and disappointed when the court returned and the Chairman said the VO's assessment was confirmed. One local surveyor said he was not surprised: the court comprised councillors from a neighbouring council and the arrangement was: 'You look after our rateable values and we'll look after yours.'

I returned to the office to report my failure to TSD. He decided to appeal to the Lands Tribunal against the LVC's decision. This we did and a few weeks later we went up to see the VO for Coventry to try to settle the matter. TSD was most persuasive in negotiation and the VO agreed to reduce the Zone A price by 2/6d.

On the train home I had somewhat mixed feelings. We had won our appeal and got approximately the right answer, but TSD had got it in a couple of hours of negotiating, whereas I had got nothing in spite of my all my efforts. I tried to comfort myself by thinking that my cross-examination of the VO might have made him think twice about giving the same evidence again before the Lands Tribunal, and convinced him of the weakness of his case. I had certainly really enjoyed conducting the appeal in Court.

Another city which was badly bombed during the 'blitz' was Southampton, the principal shopping street there then being Above Bar. One of the parades rebuilt there contained Woolworths for which Montague Evans acted – in particular, Basil (Bill) Evans, son of the firm's founder, Captain Montague Evans. Bill had a big frame, a big heart and a big capacity for drink. When the appeals for the parade were listed for hearing by the LVC, all the rating surveyors concerned, which included me, arrange to stay at the Polygon Hotel in the city the night before. We all dined together very well and retired to a room where we decided to play cards. We started on pontoon. What went mostly unnoticed during the playing of cards was a constant supply of drinks which kept arriving due to the generosity of Bill. As the evening wore on, the game changed to Slippery Sam which invariably results in somewhat uproarious gambling.

I hate to think what time we got to bed, or how much we had drunk, but we all looked and felt in a sorry state when we emerged for breakfast the next day – all except one that is. Bill

Rating Surveyors Association Guest Dinner 1949. *Top table, standing:*
H Brian Eve, President; *Top table, far end:* Douglas Trustram Eve;.
on Douglas's left: Lord Silsoe; *Front row, 2nd left:* Basil Evans. *2nd row,
1st left:* Jack Powell; *2nd Row, 3rd left (back to camera):* Eric Strathon;
Opposite Eric: Adrian Eve; *On Eric's right:* Hilary Eve.
Opposite Hilary, slightly right: Tom Dulake.

came into breakfast as bright as a new penny; he even tried in vain to carry on a conversation with one of two of us as we all drank cupfuls of coffee in a desperate attempt to wake up and feel better.

Bill was not a stupid man – indeed he was very bright and, summoning up the situation, said:

'I know my case is not listed first but would you like me to open the batting this morning and deal with the general case concerning the Zone A price for the parade? You can then follow on and put in your individual valuations based on the Zone A price we are all seeking.'

This kind offer was greeted by a chorus of assent and relief by all those present but not correct at this sorry breakfast gathering.

So Bill went first and, fresh as fresh can be, conducted our appeal, putting our general case clearly and persuasively. Then each of us in turn managed to rise from our seats and hand in our individual valuations adding:

'Finally, I would like to heartily endorse the case put before you so ably today by Mr Basil Evans.'

A vote of thanks to Bill from us all for his splendid performance; a 'hair of the dog that bit us', and we boarded the train to London and went fast asleep – except, of course, our 'friend indeed', Bill Evans, who, understandingly, quietly read the paper.

I forget the decision of the LVC, but I know either the VO or the ratepayers appealed against it, because the appeal went before the President of the Lands Tribunal, a barrister. The President went by the name of Sir William Fitzgerald QC and he was an Irishman. One of the idiosyncrasies of the Irish people, and not the only one, is that they call the ground floor the first floor, the first floor the second floor and so on all the way up to the top. What's more, they exported this irritating habit to America.

I arrived about half an hour early and the Clerk to the Court came up to me quietly and asked if I could help him with the copy of the decision that the President had handed to him the night before. He said he was a bit bewildered by it. Was it right that the first floor was valued on the zoning principle? He thought zoning and the Zone A price were applied only to the ground floor. I told him that was indeed so.

Sir William Fitzgerald MC QC, first President of the Lands Tribunal 1950–65

I read the decision and it was clear that Fitzgerald had been confused when the ratepayer's surveyor had in his evidence talked about the first floor. The President had also used the term of first floor in parts of his decision when he meant the ground floor, and in other parts when he meant the first. I went through the whole decision which seemed to have escaped the copy-editor and the sub-editor entirely. I think Sir William retired shortly after. I sorted out what the author probably meant and gave it back to the clerk who was embarrassingly grateful.

That is the only Lands Tribunal decision I have ever written – and I can't even remember whether or not I found in the ratepayer's favour.

Part V. *Gerald Eve & Co*

London W1

PARTNER

1 January 1957–1983

Chapter eighteen

Following in the Family's Footsteps

Adrian Eve (1924–1982) on his wedding day, 20 August 1949

On 1 January 1957, Bill Haddock and Billy Oliver, already salaried partners, were made equity partners, at the same time as were Geoff Powell; my older brother, Adrian; and myself. The firm's profits the previous year were £30,000 and we three young lads were each to get four per cent of it, £1200 a year. Out of that we had to pay for goodwill, which was two years purchase spread over nine years. The trouble was that the two-ninths (£267) had to be paid out of net income at a time when income tax was very high. At one time, the top rate of tax on earned income was eighty-three per cent, but on unearned income it was ninety-eight per cent.

Fortunately, with so much work on, particularly that concerning the 1956 Rating Revaluation, and more partners and staff to do it, the firm went from strength to strength. Indeed it has, with the odd slight hiccup during property slumps, never stopped going from strength to strength ever since.

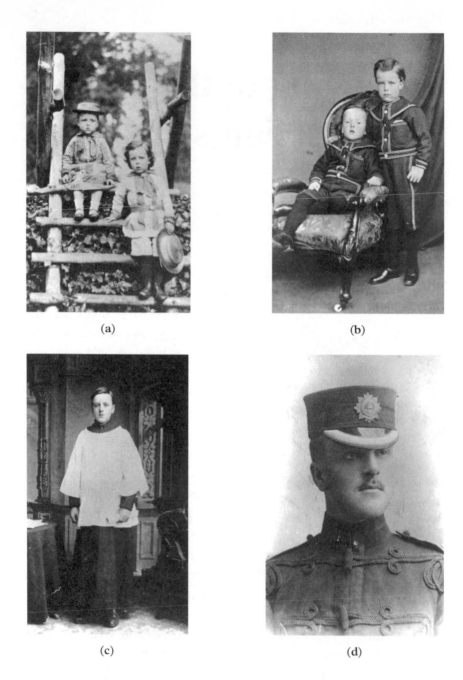

(a)

(b)

(c)

(d)

(a) and (b) Gerald Eve (left) with his brother, Frank, c. 1876 and 1878;
(c) as a choirboy, aged 15; (d) as a 2nd Lt. in 3rd Volunteer Battalion,
Bedfordshire Regiment

Even at the height of the property recession, Gerald Eve & Co.'s profits never fell by more than about a quarter. While other firm's were calling on their equity partners to contribute from their savings in order to avoid bankruptcy, or were merging with other firms, Gerald Eve & Co. sailed on alone safely through stormy seas. Today, the firm has seventy-four partners. What is the reason for the firm's continuing success? I think it is because, in good times and bad, people need advice – and in the long run the best advice is always the cheapest. The firm always aims to follow the standards and fundamental methods of its founder, Gerald Eve (see Appendix 3 for obituary).

My brother, Adrian, had originally intended to follow Uncle Stewart in his career as a physicist. It is not surprising that Adrian was inspired by his uncle's career for it was a brilliant one. Arthur Stewart Eve was the fifth child of John Richard Eve. The others were: Emily, the eldest, who had a very strong personality and whom her six brothers adored. Then there were John, known as Chum, who went to India, Herbert who I have mentioned earlier, Stewart, Frank and Gerald.

Uncle Stewart was an assistant maths master, and later Bursar, at Marlborough, and the story goes that he had an affair with the wife of a housemaster. After that I don't know whether he jumped or he was pushed, but after sixteen years there he went off to McGill University in Montreal to teach or study maths. On the other side of the landing in his office block was one Ernest Rutherford, a bluff New Zealand farm boy, who had become a distinguished physicist. They struck up an acquaintance and Stewart decided the work that Rutherford doing was far more interesting than his, so he joined him and was put to work on studying the alpha particle – something ejected by the nucleus of an atom in a spontaneous radioactive disintegration – about which I feel it would, fortunately, be out of place for me to enlarge upon here.

This chance meeting turned into a lifelong friendship. Indeed their exchange of letters for the rest of Rutherford's life was so extensive that Stewart wrote Rutherford's biography based on their correspondence. Stewart worked with him at the Cavendish Laboratory at Cambridge, together with JJ Thompson who discovered the electron.

Lord Rutherford FRS is known as the 'father of the atom' and is New Zealand's most famous son – with the possible exception, I suppose, of Sir Edmund Hillary.

When Hillary came down from Everest's summit and returned to camp, he said to Lord Hunt, leader of the expedition: 'We've knocked off the bastard.' I sometimes wonder what Rutherford said at Cambridge, when he finally managed to split the atom. Did he wander into the Common Room and say: 'Stewart, I've finally split the bastard'?

I remember Uncle Stewart as a gentle, kindly old man with sparkling, laughing blue eyes who lived at Cuttmill near Elstead, Surrey. My last memory of him was visiting him in 1945 after he had had a stroke. His wife, Auntie Betty, then told me she asked him that morning what he would like for breakfast. He struggled for a while to find the word, but then came out with:

'Four and a half minutes'.

She boiled him an egg.

Although he lived on for three years, he had his stroke shortly before the US dropped an atom bomb on Hiroshima. We think he might have known what was going to happen. His obituary is in Appendix 4.

As long as I can remember, my brother, Julian, wanted to be a doctor like Uncle Frank who, with his wife, Sally, and son, Ivor, used to come to spend Christmas with us. We used to find this uncle quite fascinating as he was a great naturalist. When we lived in the country near Loxwood, West Sussex, we would go out with him into the garden and walk down a glade. He would stop and say to us: 'Now, look at this.'

There was a spider's web.

'Now this spider will be hiding behind a leaf where he might have a larder with a fly or two he has stored up for his lunch. He will have one of his legs resting against a strand of the web so that when a fly or other insect touches it, he will sense it and come rushing out. Let's pretend we are a fly and see what happens.'

He picked up a twig. 'Watch this!'

He touched the web with the twig and sure enough out rushed the spider from behind a leaf, looking in vain for his lunch. Frank then went off in search of a real life insect and came back with one in a jar. He took off the lid and it flew into the web. The spider roared out, spat out a bit of web and twiddled the fly round and round with his legs until it was bound tight.

Although you couldn't see it, Frank told us that all spiders had venom and the spider would have given a quick bite to paralyse the fly with it. Later I remember putting some poor grasshopper on the web to see what would happen. Out roared the spider, and finding the grasshopper too big to turn, he started to run round and round the poor insect in order to bind him up. He couldn't take his prey back to his larder as it was too heavy, so presumably he had to dine alfresco that evening.

Next, Frank would go and lift up a large paving stone revealing an ants' nest with masses of eggs.

'These eggs have got to be kept at the right temperature to hatch out, so the ants will work flat out to take them down below as quickly as possible' said Frank.

He was dead right. Within a few minute all the hundreds of eggs had been taken down below.

After Frank had left, my two brothers and I decided to carry out an experiment on our own account. Frank had shown us that there were several different kinds of ants, such as large ants, small ants and red ants and we thought it would be interesting to see how they got on with each other. Perhaps, we thought, if we put them all together, we might start a war which would be fun to watch. As I write this today, this sounds rather unkind and anti-social, to say the least. You probably haven't yet got over what we did to the poor grasshopper. However, boys will be boys, so we went in search of different sorts of ants. Sure enough, we soon found a colony of large ants, one of small ants and one of red ants – which were also very small.

Many moons ago, W. Barrington Dalby, who used to give inter-round summaries on broadcast boxing matches interspersed with commentaries by Raymond Glendenning, had a favourite saying:

'A good big'un will always beat a good little 'un', and we thought it would be quite fascinating to see if that was true about ants as well as boxers. It also occurred to us that red ants, although small, had a good chance of coming out on top as in our experience they bit or stung us.

The next problem was how to get all the contestants together on one battlefield. It would be no good simply dumping one colony onto another, as they would probably just go their different ways. It would be no good building a wall round the opposing armies, as they would climb up it. Adrian, the oldest of us three, always an original thinker, came to our help – after all, that's what elder brothers are for. We would get a pane of glass and float it in the middle of the bath. It would be an island from which none could depart. The rival clans must either find a *modus vivendi* or fight to the death. A stroke of genius, we thought then; how wrong can you be?

We found a large pane of glass, filled the bath with some water and all was set for The Big Fight. Each of us brought up to the bathroom one colony of ants and deposited it on the glass. We waited quite a while as there were no ants visible, their not yet having dug themselves out after being covered with soil. We were then summoned by our mother to lunch, which meant a quick wash of the hands and down for lunch at once.

With seven of us in the family, lunches were lively affairs and we three had soon forgotten our ants. When we returned to the bathroom about three-quarters of an hour later, disaster had struck: the glass had floated to the side and the ants had climbed up the bath; hundreds of them were all over the floor going in different directions. Adrian suggested that we said nothing to anybody and that by the time our parents went upstairs again to dress for dinner the would-be warring factions would all have gone home again.

We did a quick recce to see if the coast was clear and then surreptitiously took the pane of glass back to the greenhouse. Our chief worry was that some of the ants might be termites and would make their home in the house. Then, about twenty years later, the house might suddenly fall down. Fortunately, I revisited the house a few years ago, then owned by a retired Danish brain surgeon living in tax exile. I am happy to report the house is still standing – but these things can take a long time so you never know.

Uncle Frank visited us at a time when we were all very keen on horses and ponies, and in his thank-you letter I recall that he mentioned that even the loo paper had Bronco written on each sheet. He was a great inventor: he invented a desalinator for installing in lifeboats, that converted salt water into fresh; also a contraption, I think it was called a 'haemorrhometer', that enabled you to count red and white blood corpuscles.

It was said that every time Frank's wife, Sally, took on a new maidservant and she coughed, Frank sacked the poor girl as he thought she might have tuberculosis. Perhaps Auntie Sally exaggerated this somewhat.

The story I like best is that Frank had a theory based on the fact that if a mosquito landed on you and bit you, you immediately put your hand there causing the mosquito to fly away. If however, you managed to resist that temptation – and this is the crux of the theory – then the mosquito would in his own good time inject an antidote into the hole from which he had just sucked blood so that the bite would no longer itch and there would be no swelling.

Buoyed with complete confidence in his theory and determined to prove it, when a large female mosquito (the males don't bite) landed on him and bit him, Frank resisted the temptation to swat it and let it feast upon his person for a full five minutes. Finally, weighed down with a few pints of Frank's blood, the mosquito taxied along his arm and just managed a lumbering take-off.

Did the experiment work? Frank had the biggest mosquito bite anyone's ever had, ever, ever, ever.

The lasting claim to fame of Dr Frank C Eve MD, FRCP, 1871–1952, was his new method of artificial respiration. My brother, Julian, when a student at St Bartholomew's hospital in London, was very proud when he was taught Eve's Method of Artificial Respiration and the lecturer asked if he was a relative. According to his obituary in *The Times* (see Appendix 5), *Frank's* method:

> 'has been widely adopted in Britain and America. Dr Eve
> recently modified his rocking method and showed that it was
> particularly suitable for use in open boats at sea. It has been
> officially adopted by the Royal Navy. Eve was a learned
> physician and, apart from medicine, he was a life-long
> student of general science and natural history.'

So, all in all, brother Julian as doctor, Adrian as scientist and I as surveyor each had quite an act to follow.

Chapter nineteen

'Carpe Diem' (Pick Your Day)
—Horace 65–8 BC

Brent Cross is a large shopping centre to the north of central London, next to the M25. Shortly after it opened in March 1976 I called a meeting at our office of all the other firms of rating surveyors involved. There were about twenty of them. It was decided that I would negotiate the bases of assessment on behalf of them all. They would send me details of their rents and assessments analysed in terms of Zone A.

It soon became apparent to me that the rents were well above the assessments so in no way could the rateable values be said to be above their current value. I decided to go on a fishing expedition and went to see the Valuation Officer. A recent Act (of which more later) required the VO to assess the centre at what it would have been valued at if it had been there during the year ending 1st April 1973, the date the Valuation List came into force. The VO decided, therefore, to relate it to the level of assessment in High Road, Wood Green, North London, a shopping centre in another VO's Valuation Area. However, as Peter Sellers once said in a record when he was imitating a politician orating:

'We must build, but we must build surely.'

If a boat has to be tied to a mooring, one sure way of moving the boat, but not the easiest, is to move the mooring. So off I went to the mooring – High Road, Wood Green.

I looked into the level of assessment in the High Road on behalf of all the other rating surveyors acting for retailers and found it excessive. The assessments had been increased seven-fold since the previous Valuation List in 1963 – probably more then any other shopping street in the country. I got substantial reductions in every single parade in the whole of the shopping street. I can write that pretty quickly but in fact it took an awful lot of time and effort, using whatever skill, expertise and power of advocacy I possessed.

Back I went to the VO. There was a new one by then, a lady – the only lady VO in the country. I had a female assistant, Tessa Gibbs, and the VO's assistant was an ex-Gerald Eve & Co chap, Tom Dixon. The four of us had a couple of enjoyable lunches together as a break from day-long negotiations. The VO was quite charming but took a little too long for my liking to come to her decisions, though when she did I approved of them, as you will see. For a start, because of the Wood Green reductions I had achieved, she decided to reduce substantially the one Zone A price that her predecessor had applied to every shop in the centre regardless of position.

That meant 'jam for all' but I decided some ratepayers should get more jam than others. That's life. Having agreed the top Zone A price, I was anxious to prove that other parts of the centre were less valuable.

On inspecting the place, I noted that a very clever use had been made of the difference in levels. On the south side you drove in at ground floor level and parked as near to the middle as possible. As the surface car park filled up, cars had to park further and further towards either end. On the other side was a multi-storey car park which shoppers tended to use only when the surface car park was full. (Shoppers, particularly women, prefer surface car parks to multi-storey ones.) The entrance to the shopping centre was at first floor level only, but here again there were entrances at either end which were used as the multi-storey car park filled up.

It became clear to me that at its busiest – say Saturday lunch time – the centre had been designed so that pedestrians were pretty evenly spread from one end of the malls to the other and from one shopping level to another.

My plan of campaign was to take a pedestrian count on the least busy day so that the upward and outward spread of pedestrians was at its minimum. Normally this day might be a Monday but in this part of the world where there was a large Jewish community, Monday was quite a busy day. Orthodox Jews were not meant to use their cars on Saturday, the Jewish Sabbath, so did little shopping, and the shopping centre was closed on Sunday. That made Monday quite busy, so Tuesday was the day I chose.

I gathered together a fair number of staff and positioned them at strategic points in all the shopping malls, each person armed with a clipboard and a 'clicker' to count pedestrians. In true Army fashion, we had synchronised our watches and at H-hour on D-day the count began, with suitable breaks for coffee, lunch and tea. The task force had been told that if someone accosted them they were not to reply, but simply hold up their clicker and say '53...54...55...' and so on. The accoster would then get the hint and move on. One very pretty girl in our team, Tessa Gibbs, found she had to do this quite a lot, but it worked eventually, even if she had to continue counting out loud rather longer than the rest of us: '53...54...55...56...57...58...59...60...61...62....' and eventually her admirer would give up trying to chat her up and walk off.

The results of the pedestrian count were absolutely splendid – there were massive differences between the flows in the middle and those at either end of the malls, and the first floor flow was substantially less than that on the ground.

To illustrate the result graphically, I constructed a pedestrian flow diagram that looked rather like a London Underground Map, except that the width of the lines was varied to accord with the amount of the flow. Armed with this, I went to see the VO.

Taking care to make no mention of the pedestrian count, the first thing I did was to agree with her in detail the basis of assessment, namely the Zone A price, for the peak trading position which was in the middle stretch of the ground floor mall. That was a very important first step because it meant I had established a maximum zone A price. To make absolutely sure there could be no going back, I agreed to the exact pound the assessments of several shops where I was acting which were in the peak position.

Next I explained to the lady that unlike rents passing in a High Street, these had been entered into with the traders having no experience of their relative trading positions. Normally, when a shopkeeper is deciding how much to pay for a shop, he goes and looks at the position on a busy day, probably Saturday, to see how busy it is. If there is a choice of shops he may take a pedestrian count to find which pitch is the busiest. At Brent Cross he could do none of this.

Whereas the trader could not be wise after the event – he was stuck with the lease – we could and should be wise when determining the various rateable values throughout the centre.

I will stop the action here for a moment to explain about the definition of 'Rateable Value'.

Whereas other rating surveyors might take with them, to negotiation meetings or to court, heavy books on rating by learned authors which contained a mass of case law (such as *Ryde on Rating*), I was unable to. I had for some years been suffering from a 'slipped disc' and found it painful to carry heavy weights. I decided therefore that I would simply learn off by heart the statutory definition of Rateable Value. To hell with all the case law – after all it was no more than the result of applying common sense to the wording of the statutes. To arrive at 'rateable value' for shops, offices etc, but not industrial properties, you first had to decide on the 'gross value' and then deduct a set amount laid down by statute to allow for the cost of repairs etc.

So I learnt off by heart that Section 68 of the Local Government Act 1948 defined 'Gross Value' as meaning

> ... the rent at which the hereditament might reasonably be
> expected to let from year to year if the tenant undertook to
> pay all usual tenant's rates and taxes and the landlord under-
> took to bear the cost of the repairs and insurance and the
> other expenses, if any, necessary to maintain the hereditament
> in a state to command that rent.

Having just written that, I have now checked it with the wording in the Act and it is word perfect. It is not a lot for a rating surveyor to learn and I commend doing so for anyone with any aspirations in that direction. It's all you need to know really – that plus a bit of nous.

Back to the negotiating table and, with a few more meetings and a lunch or two – she had a reputation for being rather slow to come to decisions – I managed to get the VO to agree to reductions, some substantial, everywhere in the shopping centre, over and above what she had already given for the peak position.

I was more than satisfied that I had been able to agree a proper pattern of values in the various parts of Brent Cross Shopping Centre which fairly reflected their respective pedestrian flows – on a wet Tuesday in February.

★★★

Some years later, I conducted pedestrian counts for the same purpose in other shopping centres, such as Milton Keynes, but there all the car parks were on the surface and were well spread to ensure the relative pedestrian flows were fairly similar in peak and off-peak shopping hours.

The most difficult shopping centre to deal with in this respect was a large new one at Sutton Coldfield. To our great disappointment and consternation we found that the pedestrian flow in the principal shopping mall at ground level was similar to that on the upper level. We couldn't think how to justify a reduction. I called in some really top-class brains – there were always plenty to call upon in Gerald Eve & Co. – and one of these had a Ph.D. I threw at them all the statistics of the pedestrian counts and however hard we all scratched our heads we could think of nothing.

Almost as a joke, I said:

'Upstairs, we have added together the flows either side of the wells going along the middle of the mall. Why not rehash the chart and show the flows separately? Then we could halve the Zone A price'

'Because they'd never wear it', said one of the super-brains, 'That's why.'

'Well, have either of you got any brighter ideas – or indeed any ideas at all?' I said.

Silence.

'Right then, that's what we are going to do. We've nothing to lose. The flow of people walking either side of the ground level mall is counted as one because there's no obstruction in between. They can wander from side to side window-gazing, but the mall upstairs is really two one-sided malls with only the occasional passage joining the two. And in any case, it is well known that one-sided shopping streets are never much good. The exception that proves the rule is Princes Street, Edinburgh. We'll count the pedestrian flows for each side separately.'

Consequently, we altered all the pedestrian flow diagrams and reduced substantially all our valuations for shops on the main malls on the upper level.

This approach was accepted, without comment, by the VO and by the valuer acting for the local authority. I was beginning to realise I was an 'ideas man', a creative thinker, and that I needed bright people around me with both feet on the ground who could evaluate my ideas and, if they were any good, knock them into shape. Later on in my career, I was rather pleased that I had invented a form for rental analysis which was called HME Mk. 15. The HME was my initials and the fifteen meant that the staff in my department, including myself, had found fourteen ways of improving it.

The important thing was to create an atmosphere in my department where even the lowliest was encouraged to think up ideas – and never to ridicule anyone who made a suggestion even if it was a silly one. Indeed when we got together to find our way round a problem I used to start the ball rolling by making a really silly suggestion myself. I found it came fairly easily.

An excellent example of this was an occasion when we had to create about a thousand new files – one for each shop assessment we had been asked to deal with. The normal file had written on it by hand the client's name and address, when and by whom we had been instructed, e.g. solicitors, and their name and address. It was substantial, being made of thick card.

I called a meeting of my department to decide what sort of file cover to have and what to write on it. There were junior partners, associates, chartered surveyors and chaps fresh from college or University. My first question was: 'Do we need file covers at all?'

A student said he thought it preferable – otherwise when you opened a drawer of a filing cabinet there would be just a mass of papers and you might have to look through them all to find the one you wanted. I thought he had made a good point and told him so. He was delighted.

By this method we probably saved a large amount of time and money. By devising a new system of numbering and abbreviations, the writing on each file was reduced by about 90%. The cost of the file cover was also reduced by about 90%.

'Why not have a really cheap, flimsy file cover?' I asked.

'Because some of them, the more heavily used, would fall to pieces after a while,' said one.

'That would only happen to a few and then you could replace them with a new one' said another. 'It would still be miles cheaper than having more expensive ones for all the files.'

He was dead right and that is what we did.

Gerald Eve had introduced this method of 'brainstorming' into the firm, and I have written earlier that Bill Haddock told me that Gerald would start the meeting by saying:

'Gentlemen, let us all air our ignorance.'

This brainstorming was to prove invaluable to me when I gave evidence before the Lands Tribunal in the Siggs case which concerned valuations for leasehold enfranchisement, but that is a story I will keep for later; it was probably my most successful case before the Lands Tribunal, so if I save it till near the end of this book you might remember it better.

Chapter twenty

Two Sellers and a Buyer

I remember someone once criticising one of David Niven's very amusing books of memoirs because the author indulged in namedropping. I have never understood this criticism, because David Niven always seemed to me to be a very modest sort of person who would not seek self-aggrandisement in this way. In any case, he went to the same school as me and served in the same regiment in the army, The Rifle Brigade, so I am jolly well going to come to his defence against such an accusation. Surely, if you have met famous people it is much more interesting, other things being equal, to tell stories about them rather than about people your readers have never heard of – however worthy.

After giving this matter much thought, I have decided to tell you about the famous people I have met in my life only if there is an interesting or entertaining story to tell about them. So I will not tell you about the time I danced next to Princess Elizabeth at a Charity Ball, however exciting I thought it was at the time, because that is the whole story – except that she was dancing with Lord Louis Mountbatten and that, when they went to sit down at their table she and Princess Margaret got out their powder compacts and lipsticks and made up *in public*.

Nor will I tell you about when Sir Hartley Shawcross QC MP, the Attorney General, attempted to sneak to the front of a ticket queue I was in at a London theatre and I shouted out to him: 'Excuse me, I know you are the Attorney General, but does that entitle you to barge to the front of this queue?'

He decided it didn't.

The first famous person to qualify under the criterion outlined above is the inimitable, the one-and–only Mr Peter Sellers.

The partners of Gerald Eve & Co. in my time made an effort not to specialise too much in one sort or property or one sort of advice, so we all liked to deal with shops, offices, factories and residential properties, etc.

Peter Sellers and Britt Ekland outside their house in Elstead, Surrey

We also liked to do rating, compensation, landlord and tenant and other sorts of valuations. In addition, we liked to advise on town and country planning matters. Once such job came to me in 1965 via Peter Sellers' financial adviser. This adviser was kind enough to ask me to lunch at the Ritz Grill to explain his client's problem. (If only all clients did this.)

Peter Sellers owned a house at Elstead, Surrey, near the Hog's Back. He wanted to house a married couple he wished to employ and had put in an application for planning permission to build a bungalow in the grounds. This had been refused. Would I appeal against the decision and give evidence at a local public enquiry? I said I would advise him on the matter.

I went to inspect the house which was a large, old, black–and-white Tudor cottage which had been much extended. To my great disappointment neither Mr Sellers nor his wife, Britt Ekland, were in residence. (Britt was his second wife, 1964–8, of four.) I went to see where it was proposed to build the bungalow. It was on a spot well away from the house. The property was in the Green Belt and in an Area of Outstanding Natural Beauty, so there was not a hope in hell of getting planning permission from the local planning authority or the Minister on appeal. I rang Sellers' adviser and told him so.

'However,' I said, 'if you put in an application to build on to the house in an appropriate style there is every chance of it being permitted. What you want is a good architect, and if you like I will find a suitable one for you.'

'All right,' he said 'but you had better be very careful.'

'Why do you say that?' I asked. 'Of course I will.'

'Because the last architect he employed ran off with his previous wife.'

I made a mental note not to choose anyone I thought Britt Ekland might fancy.

I soon heard of an architect who specialised in extending Tudor houses and went to meet him. Although good-looking, I felt he was a little too old for Britt so accepted his offer to take me to see one of the Tudor houses he had extended. When I saw the house, I told him I thought the extension was 'not bad' but I preferred the original Tudor cottage.

'What you call the original Tudor Cottage is the extension I put on', he said. 'The 'not bad' bit is the original Tudor cottage'

I gave him the job (wouldn't you?) and we went to have another look at Peter Sellers' house. As we walked round the outside and came to what seemed like the oldest part, he stopped and said: 'Ah, yes. I built that bit on in 1935. Give me a little time while I go and track down some derelict farm buildings from which I can obtain suitable building materials.'

I made an appointment to see Peter Sellers and on my arrival was told he was in the garden doing a television appearance for the British Heart Foundation – he had recently, in 1964, had a serious heart attack. He dropped everything and came into see me, receiving me with the utmost courtesy. I explained to him what the solution was to his problems and I noticed that as he talked to me his accent kept going very slightly like one of the many characters he imitated on the radio and in films; and then he would quickly pull himself back again to a normal voice, thinking: I mustn't fool around. This is a serious matter I am discussing with my planning consultant.

When I read his biography after his death, his biographer wrote that Peter Sellers did indeed have trouble being himself, and sought refuge from this by becoming one of his many characters he imitated.

Over the next few months, I visited the house several times to see how the extension was shaping up and to deal with his rating assessment. I discovered that if I wanted to meet the ravishing Britt I had to come in the afternoon, as she did not emerge until midday. She was every bit as beautiful as she looked in films. So was her figure.

Later, Peter's adviser asked one of my partners, Michael Hopper, to make a valuation of the house because Peter, who was away abroad on holiday, was thinking of selling the property. Michael made what I thought was quite a full valuation of the house but when Peter came back it transpired that he had already negotiated the sale of the property for a price way above this valuation. He had agreed to sell it to someone he met on holiday who had never even seen the place.

'Who was the buyer?' I asked. Came the reply:

'Ringo Starr.'

A Liverpool drummer negotiating with a very shrewd Jew from London's East End? Do me a favour. That's not negotiating. That's taking candy from a kid.

Michael Hopper, partner in Gerald Eve & Co and, later, a Member of the Lands Tribunal 1990–9

Chapter twenty-one

Bribes, Prompts, and Perry Mason

I used to do quite a lot of Town and Country Planning work. Its advantage was that it got you out into some beautiful parts of the country – often those parts that builders wanted to develop against the wishes of the local planning authority and the 'Nimbies' (not in my back yard).

Nimbies could be pretty frightening. Once I gave expert evidence at a public local inquiry concerning an appeal against refusal to give planning permission to build houses on some land near the sea at Ferring, near Worthing, Sussex. My assistant in that case was a young girl newly graduated from university, Kathy Sanders, and, as we got near to the hall where the inquiry was to take place, she went pale when we came across droves of obviously angry people walking towards the venue. I smiled to reassure her and told her there was nothing to worry about, but my inner feelings were very different.

As we entered the hall, which was already packed, we heard loud mutterings and epithets like 'scandalous, disgusting, appalling'. Half the people looked like fierce, retired colonels with red faces, military moustaches and tweeds. The other half were their equally formidable-looking 'other halves'. All the world and his wife had come to the inquiry as objectors and were going to get up and say their pieces in no uncertain terms. 'Disgusted' from Tunbridge Wells seemed a pretty mild-mannered sort of chap compared with 'Ferocious' from Ferring.

When Kathy and I took our places in the 'hot seats' and the objectors realised that I was the enemy, the glares and mutterings increased. Fortunately, the Planning Inspector was also a retired Army officer, and so at the slightest outbreak of barracking during my evidence, such as cries of 'rubbish' and 'nonsense', he clamped on it with an iron hand:

'All objectors will have their say in due course. In the meantime, I must again ask you not to interrupt the evidence.' Fun it was not.

Then there was the enquiry into the proposed central area development at Sevenoaks, Kent. We were acting for the Council, and the Surveyor and Engineer had formally submitted the plan for the new road system in the central area. It was time for cross-examination by counsel for the objectors.

'Mr Smith, as I understand it, you are entirely happy that this new highway system which you put forward for the central area will work well and that the objections to it are wholly unsubstantiated?'

'No!' said Mr Smith

'I am sorry, I didn't catch that,' said counsel

'I said, "No, I'm not!"'

Barristers are trained never to ask a witness a question unless they are sure of the answer he or she will give, so counsel was taken aback – especially at this wholly unexpected windfall 'But, Mr Smith, surely this Highway Plan is yours.'

'No, it's not. My Deputy got it out when I was on sick leave and I do not approve of it.'

Our counsel scribbled a note and it was passed to Billy Oliver. It said

'What do we do now?'

Billy scribbled on the note and it was passed back to counsel. He opened it. Billy had written:

'Send for Perry Mason.'

★★★

There was some more light relief, I recall, at a public local inquiry in Sussex. There were a number of objectors to the development proposed by our clients, and the inspector had in front of him a large-scale map. It had been prepared by the local planning authority and showed the land the subject of the appeal and the names of the houses in the immediate neighbourhood.

'Are there any more objectors?' asked the Inspector.

A man stood up: 'Yes, Sir. The wife and I'

'Thank you. Where do you live?'

'Right opposite, Sir'

'What, in this house called *Navarone*?'

'Yes, Sir.'

'What are your names, please?'

'Mr and Mrs Gunn.'

★★★

One of the things I did not like about giving evidence at planning inquiries was that there was such a lot of bumf to read about Ministerial planning policies. I am a very slow reader and it was quite impossible for me to get through the vast amount of literature generated by what was then called the Ministry of Town and Country Planning, much of which was relevant to the issues raised at a public local inquiry. The job of the Planning Inspector was to advise the Minister whether the development proposed, or in some cases the Development Plan, accorded to ministerial policy. My job as expert witness was to show that it did so; thus I had to wade through all the bumf to prepare my proof of evidence. This was in stark contrast to rating where all I needed to know was Sec 68(1) of the Local Government Act 1948 off by heart.

The biggest inquiry I attended was that into the objections to the Surrey Development Plan which was held at County Hall, Kingston-on-Thames. It was a huge hall, absolutely crammed, and I noticed many of the top planning QCs in attendance. I was feeling pretty nervous at having to give evidence knowing that the County Council's QC was a formidable cross-examiner.

Whilst waiting, I met my cousin, John Richard Eve, who was also to give expert evidence, and we had a chat. When the Inspector opened the inquiry I was horrified to hear that my client's objection was to be taken first and so I would be the opening batsman.

Our counsel opened our case and then called me to the witness box to read my proof of evidence. When I had done that the County's QC got up to cross-examine me. He asked me one question which I answered by saying what I thought was the Government's policy on that particular aspect as expressed in some rather lengthy booklet on Ministerial planning policy.

'Where does it say that?' he asked.

I had absolutely no idea but I thought I had read it somewhere. However, I could hardly say:

'I've absolutely no idea; but I think I've read it somewhere.'

I felt my client in the audience would think: Is this the guy I paid all this money for so as to give a pathetic answer like that?

I opened the booklet. The passage might be anywhere – I wouldn't find it in a month of Sundays. Panic set in. Suddenly I heard a voice in the front row on the very right, just next to the witness box that I was in, say very softly:

'Page 27, paragraph 4.'

Manna from heaven – if it was correct. I had nothing to lose.

'I think' I said 'you'll find the passage to which I am referring at page 27 of the booklet. Yes, here it is at the fourth paragraph. Shall I read it out to you?

'No, Mr Eve, there's no need for that. Moving on to another matter...'

I looked up to see who my prompter was. It was my cousin, John. He gave me a wink.

Every time I've seen him since, I buy him a drink.

★★★

Sometimes your successes don't seem to benefit your client as much as you hoped for. One such case concerned some fifteen acres of wasteland not far from London Airport. The local planning authority had refused permission for housing development and I was the expert witness in the resultant appeal. A family who were slaughtermen owned the land and they were all unable to read or write.

We won the appeal, which meant that the land was worth an absolute fortune. I heard later that the owners, on hearing the result of the appeal, invited the entire neighbourhood to a party that went on non-stop for a week. Moreover, I heard that they had never been sober since.

We had a gentleman's agreement that if we won that appeal, we would sell the land to the local authority at open market value. The day after the result of the appeal was published in the local paper, a man from a firm of house-builders telephoned me at the office to say that they would like to buy the land. I explained about the gentleman's agreement. Undeterred, and presumably hoping I would not behave like a gentleman, the man said that if we were to sell the land to them they would instruct us to sell all the houses (about 150 of them). I explained that we were not house agents and therefore did not sell houses. Ever resourceful, he went on to say that he would give us an amount equal to the total selling commission for the 150 houses in any case, even though he had to instruct some else to sell them. I said we could not accept such an offer. He finally said:

'Well, if your firm can't accept that money then I will make it payable to you *personally.*'

I declined that also. It was not till twenty years later that I took the trouble to work out just how big a bribe I had been offered. It amounted to about £30,000 which today, allowing for inflation, would be over £200,000.

Chapter twenty-two

When is a Flood Plain?

This planning appeal concerned some land near Sunbury-on-Thames. A firm of builders had had planning permission for residential development refused on the grounds that the land was subject to flooding. I went to the Council Offices and they showed me the Flood Map they had prepared. There had been some exceptionally big flooding in 1947 and this map showed coloured blue the land that had flooded on that occasion. It was rather a small-scale map and the blue had been overprinted on it. Certainly, my clients' land was just within the edge of the blue, but overprinting is not always that accurate. I thought it was worth further investigation and, in any case, unless I could disprove the allegation of the land being flooded in 1947, there was absolutely no hope of winning the appeal.

I went back to the office and did a bit of research as to where I could get some aerial photographs of the appeal site. I soon tracked down a firm that made aerial surveys and on visiting their offices found a photograph of the 1947 floods that included the appeal site. The trouble was that the site was in the background and it was not immediately apparent whether or not it was flooded. I asked if there was an expert on interpreting aerial photographs and they produced him. I asked him whether in his view the appeal site was covered in water. He produced a large magnifying glass and pored over the photograph for some considerable time while I waited impatiently. Then he stood up, turned to me and said:

'Frankly, I can't answer your question definitely one way or the other.'

I thanked him and off he went. That was a fat lot of good. He had left the magnifying glass behind so I picked it up to have another shufti at the photograph. I looked at the land in the foreground and middleground that was clearly flooded. Although it was a black-and–white photograph you could tell it was covered with water because the land was all one shade. It was also lighter than the unflooded fields because the water reflected the white clouds. I studied the appeal site again. It was definitely very slightly darker than the obviously flooded fields and had a faint texture on it. That was good enough for me. I bought the aerial photograph at considerable expense and told the clients I would give evidence in support of the appeal.

At the enquiry I produced the aerial photograph and explained to the Planning Inspector why I thought the land had not flooded in 1947. He listened patiently. When I had finished explaining, I was cross-examined by the Council's advocate:

'Mr Eve, you will appreciate I am sure, will you not, that the interpretation of aerial photographs is a highly skilled and specialised profession?'

'Yes.'

'Have you had any training in it all?'

'No.'

The advocate stood there motionless for a few moments to let my answer ring round the room for maximum effect. Next, he looked round towards the Inspector in triumph:

'No more questions.' Then he sat down quickly.

My clients shifted uneasily in their seats. Their prospects of making a fortune out of this land seemed to be pretty close to nil. I looked for the nearest hole into which I might disappear.

The Council's advocate called the Planning Officer who outlined the Council's, and indeed the Government's, policy on not allowing building on land subject to flooding. The criterion for this was the extent of the 1947 floods.

The next witness was the Flood Officer. I forget his official job description; however, he was not the Flood Officer himself because he was on leave; he was deputising for him. He seemed to be a charming old boy with a distinctly rural appearance and accent. He was taken through his evidence-in-chief, produced the Flood Map and confirmed that the appeal property was shown on the map as being within the flooded area.

The cross-examination went something like this:

'Mr Brown, this Flood Map is drawn to rather a small scale. Can you be sure that the appeal site is within the flooded area shown in blue on the Flood Map?'

'If you mean: " is the appeal site coloured blue on the Map?" Yes it is. But if you are asking me if it actually flooded, well, that's another matter.'

'Are you suggesting, Mr Brown that you cannot be certain the appeal site flooded?'

'No, I am not. I'm not uncertain at all. I'm quite certain what flooded and what didn't because I took myself there at the time and made a note of what did flood and what didn't. Look, this Flood Map is not accurate round this part of the river. Do you want to know what really happened?'

Our barrister did his best to conceal his growing excitement and put on as casual a tone as he could manage as he said:

'I would be much obliged, Mr Brown; perhaps you would be kind enough to take the Ordnance Survey map showing the appeal property over to the Inspector's table and show him which fields to your own personal knowledge flooded and which didn't.'

Mr Brown picked up the map and went over to the Inspector who handed him a red biro and said:

'Mr Brown, please mark on the Map with this biro the boundary of the flooding in the neighbourhood of the appeal property which you found on your inspection at the time.'

I hurried over to the Inspector's table with my copy of the O.S. map so that I could copy the line that Mr Brown was going to draw. Mr Brown picked up the biro and drew a line showing the boundary of the flooding. It was well clear of the appeal site.

I turned round to my clients and put my thumb up. Even though we had to wait two or three months for the Inspector's Report and the Minister's decision, I knew we had won the appeal.

And so it proved.

Chapter twenty-three

Difficult Questions

I have mentioned, as you would expect, two of my most successful planning appeals but, as I will shortly relate, they didn't all go quite so swimmingly.

My partner, the late Tom Dulake, was in such demand as an expert witness on planning and valuation matters that a number of us assistant chartered surveyors spent much of our time writing proofs for him. It was a joy to watch him in the box make the most of the extensive research and effort that we had put into preparing these proofs. His intellect, character, presence of mind and coolness under fire in the witness box were remarkable – as was his attention to detail.

On one occasion I had written a proof for TSD giving evidence before the County Court in Northampton concerning what rent should be payable on renewal of the lease of a corner shop. The road on which the shop had a return frontage was not a shopping street and so I did not bother to go down it, but TSD was asked by the judge where that road went to and what sort of property fronted it.

Oh, lord! I thought; I should have found that out and included it in the proof of evidence. TSD didn't bat an eyelid:

'There are terraced houses on both sides throughout its length, apart from a small repair garage on the left about three hundred yards away round the corner.'

I was flabbergasted but there and then I made a firm resolution: whenever I went into the witness box in future I would always go that bit further, and always learn that bit more about the subject, than seemed strictly necessary.

Unfortunately, I was unable to apply that maxim some years later when my partner, Billy Oliver, fell ill and I was asked to stand in for him at a planning enquiry the next day. I was given his proof of evidence which had been prepared by his assistant and I had time to inspect the appeal property that day. I read the proof and it seemed all right to me.

The client was a very colourful, extravert character who owned a chain of garages in South London, and went around with a large roll of banknotes in his pocket and a couple of minders. He had bought an ugly, rambling, old house in the country which he wanted to demolish and replace with a block of flats. If he was refused permission, he said he would convert the building into a hostel for homeless West Indian immigrants. Failing that, he would turn it into a pig farm.

The opposing counsel was Mr Vaughan Neil, a very able barrister, who was later retained by a firm of house-builders, to act for them in all their planning appeals.

As it soon became clear, Mr Vaughan Neil had studied the career of Norman Birkett, a famous barrister who, before he became a judge, earned a reputation for getting alleged murderers proved innocent. At least, his clients were found 'not guilty' but whether or not they were innocent is another matter. To obtain an acquittal, counsel had to do no more than put a 'reasonable doubt' into the minds of the jury – and Birkett was pretty good at doing that.

The case to which I refer was called the Rowse Murder Case and concerned a person who was found dead in a burnt-out car. The prosecution had called a metallurgist as an expert witness. As I have mentioned earlier, the expert witness is the only witness who can give an opinion, and the task of opposing counsel is to show that the witness's opinion is of little or no value. He does that by first trying to throw doubt on or disprove the expert's competence.

In this case, I read somewhere that Birkett took a couple of months to decide what his first question to this expert witness would be. It was this:

'What is the coefficient of expansion of brass?'

'I don't know.'

'What is the coefficient of expansion of copper?'

'I don't know.'

'What is the coefficient of expansion...' and so on.

The expert witness, confident and assured up till then, had his confidence shattered by these questions – and even more so when Birkett went on to demonstrate that the knowledge of these coefficients was of vital importance to the case. He got his client acquitted.

In my case, the clients had proposed to plant a clump of trees in order to obscure the development from the neighbours. Mr Vaughan Neil's first question to me in cross-examination was:

'What is the average rate of growth of holly per annum?'

'I don't know.'

'Don't you have any idea, Mr Eve?'

'No.

'What is the average rate of growth of beech per annum?'

'I don't know.' This was awful, so I thought I'd better natter on a bit:

'I suppose it would depend on the weather in any particular year, the amount of rain there was, and sunshine, and the average temperature – that sort of thing.'

'Obviously, .Mr Eve, but you have no idea of the *average* rate of growth per annum of holly or beech, or I suspect any other tree, have you, Mr Eve?'

'No, that is true, and I'm not going to guess.'

Mr Vaughan Neil then explained to the inspector the relevance of these questions, because the appellants proposed to plant a clump of trees, which they said would conceal the proposed development.

I felt an absolute fool and whilst driving home thought of some magnificent answers I should have given in cross-examination, but then I always did think of magnificent answers to give in cross-examination – on the way home.

It didn't help my feelings when the client, obviously a keen gardener, said he'd wished counsel had asked *him* those questions about the trees because he knew the average rates of growth of all of them. The next day, when I got back to the office and went to see my partner, Bill Haddock, for a bit of what they call these days victim support and counselling, I told him the questions. Bill said:

'A good rule of thumb, Hilary, is one foot for hardwoods and two feet for softwoods.'

I pass on that tip gratis to anyone reading this who may find themselves in a similar position to me.

I might as well get off my chest now another acutely embarrassing moment in the witness box. Then perhaps I will feel a little better about it.

I was giving evidence at a planning enquiry concerning land upon which my client wanted to build a 'motel'. My counsel was a very distinguished one, Jack Ramsay Willis QC, who later went on to become Sir John Willis, a High Court Judge. He was quite brilliant and had a wonderful intellect. I was amused when he told me his favourite occupation on a Sunday afternoon was to watch a really good Western on the television. Me, too. We loved the fact that in a Western you always knew where you were with everybody, because the 'baddies' wore black hats and the 'goodies' wore white ones. It would, we agreed, make life so much easier for the juries to give their verdicts if the same rules applied in the Criminal Courts.

But to get back to my motel, there is, or was, a planning regulation called the Use Classes Order. This Order grouped together a number of uses into one class and said you could change from one use to the other in that class without needing to obtain planning permission. For instance, you could change from a shoe shop into a clothes shop without planning permission because they were in the same Use Class; but a restaurant, for instance, was in a different Use Class and thus you needed permission.

The opposition comprised Mr Thomas Sharp who was a unique one-man-band. He acted both as an advocate and as an expert witness on planning matters, so you got two for the price of one – or perhaps he charged double the price. He was also an author of books on Town and Country Planning. The barristers didn't like him because he was doing their job, and the planners didn't like him because he was a planner cross-examining his fellow planners in the witness box, one of his tasks being to pooh-pooh their professional competence.

His first question to me was:

'Mr, Eve, you call yourself a planning expert, do you not?'

I had to answer 'yes'.

'In what Use Class is a "motel" in the Use Classes Order?'

'Er, I'm not sure.'

'You've no idea?'

'No'

'But nonetheless you call yourself a planning expert.'

'Yes'

'You see, Mr Eve, this is extremely important in this appeal because if it is allowed, which God forbid, your client could change the use of this property to any other use in the same Use Class without the need to obtain planning permission.'

Jack Willis was on his feet. The man in the white hat had pushed open the saloon doors just in time and there was no one quicker on the draw than him. He shot straight from the hip:

'Mr Sharp, I see you have the Use Classes Order in your hand. I think it would be helpful to the Inspector if you would be kind enough to pass the Order to Mr Eve. Then we will be able to have the benefit of his expert opinion as to in which Use Class this motor-hotel might be included.'

That is what you pay good counsel for.

At the beginning of a proof of evidence for a planning appeal, under 'Qualifications' we always put this sentence:

'I am familiar with the appeal property and the surrounding area.'

After the hearing of one public local inquiry the inspector, not having a car available, suggested that either the Planning Officer or I drove him to the appeal property where the three of us would make a joint inspection. I volunteered to take him and he accepted. I thought I knew the way but I soon became hopelessly lost. I got more and more embarrassed and the inspector got more and more impatient. He must have been thinking: this witness doesn't seem to be at all 'familiar with the appeal property and the surrounding area'.

Perhaps fortunately, I cannot remember whether or not we won that appeal.

Soon afterwards I was giving evidence in another planning appeal concerning some land near Swanage, Dorset. The Council's advocate started his cross-examination by asking:

'Mr Eve, when did you first visit the area?'

'1936.'

Everyone, except counsel, laughed. My parents had taken the family there in that year.

Counsel would have done well to remember the old adage: Never ask a question of a witness unless you think you know the answer.

After he had recovered from my evidently amusing answer, he asked me when I had first inspected the property in connection with this appeal. I told him it was a couple of months ago.

He then asked me when I had next inspected the property and I said:

'Last night.' He then said:

'You mean in the evening when it was dark?'

'Yes,' I said.

'That was a funny time to inspect the appeal property was it not?'

I forget my answer – it was probably that I did it to refresh my memory, but the real answer should have been:

'I didn't want to look a proper Charlie and lose my way again should the inspector ask me to drive him to the appeal property.'

One counsel I had, Bill Scrivens, was quite brilliant but he had become a very heavy drinker. He drank much too much in our hotel the night before the inquiry and, to avoid him drinking any more, I suggested we retired to our rooms. He declined to do so and I, unwisely, left him in the lounge with the waiter hovering to supply him with more drinks. There was no sign of Bill at breakfast and when I got to the venue for the inquiry there was no sign of him there either. Then one minute before the inquiry was due to open, the door opened, a strong whiff of alcohol entered the hall, shortly followed by a very red-faced Bill.

The inspector opened the inquiry and called upon Bill to open the proceedings. He rose to his feet and we all waited to hear the customary opening address by counsel. He took in a deep breath and with some effort and said:

'Without more ado I will call Mr Eve as a witness to read his proof of evidence'– and sat down again.

My reading the proof, together with my cross-examination by opposing counsel, and answering questions from the inspector, took up all morning. Bill had a drink at lunchtime and then I must admit, made a brilliant cross-examination of the Planning Officer.

It was he who told me that when opposing counsel puts to you his final multi-faceted all-embracing question and sits down, you don't make a long rambling answer trying to deal with all the points he has raised. You just give a very short denial such as 'Not at all'. Here is an example:

'I put it to you that the whole of your evidence is a tangled tissue of lies deliberately concocted to discredit my client.'

'Not at all.'

Bill was a very talented barrister with a good, quick brain, but at that stage of his career he was a little unnerving to have on your side.

I was not prepared to give evidence in planning appeals if I did not think they had a reasonable chance of success. This meant that I had to be able to make a reasonable case that the proposed development was in accordance with Ministerial planning policy.

I remember going to inspect one appeal site in the company of my assistant who was fresh from university. The client, a builder, wanted to build houses on the land. After inspecting the site and the surrounding area and studying the relevant plans and other documents, I told my assistant that I would not give evidence in this case. He said he was surprised at my decision because I would be forgoing a big fee if I did not give evidence. I said I would get my reward in heaven, but went on to explain that it was not quite such a high-minded, altruistic thing to do as he might think: in the long run, you will make more money by selecting your cases rather than by taking them all, for you cannot build up a clientele if you lose most of your appeals.

I telephone the client there and then and told him that I did not think the appeal was worth pursuing to an inquiry. He accepted my advice. When I got back to the office with my assistant the phone rang. It was the client again. The County Council wanted to buy a large area of my client's housing land on which to build a Comprehensive School. Would I act for them in negotiating the compensation payable? I accepted. It was an easy job and the fee, on a fixed scale payable by the

Council, would be several thousand pounds. I turned to my assistant and said:

'I feel I should point out to you, Donald, that you don't always get rewarded for your professional integrity quite as quickly as that.'

A few more memories concerning planning appeals come to mind. The first concerned a public local inquiry into a proposal by Victor Value, a firm of grocers since taken over by Tesco. Victor Value proposed to demolish the Newcastle Arms, a pub in Weybridge High Street, in order to erect a supermarket. One of the objectors, a Commander Church, got up, said his piece and concluded with the memorable remark:

'I didn't fight in two World Wars for my local pub to be turned into a supermarket.'

Whilst we should all have been very grateful to him for fighting in two World Wars – two more than I have fought in – I do suggest, with the greatest of respect to him, that none of us there did actually think that that was precisely the purpose for which he had fought – in two World Wars, bless him.

Another memory I have is of a visit to the office by Mr Frank Taylor who I think was one of the founder members of the building contractors, Taylor, Woodrow.

Like Peter Sellers, he owned some land near Elstead, Surrey, near the Hog's Back, designated as being in an Area of Outstanding Natural Beauty. And it was in the Green Belt. He wanted to erect a dwelling on it. Could I help him? (Could I hell!) He showed me the planning application and the refusal and said he had written to the following people: The Queen, The Archbishop of Canterbury, The Prime Minster, The Minister of Town and Country Planning, his Member of Parliament, the Leader of the Opposition, the Chairman of the Surrey County Council, the Chairman of the District Council and the County Planning Officer. Was there anybody else that I thought he should write to?

I scratched my head and nearly said: Uncle Tom Cobley, but didn't feel somehow that he would think my remark funny. I fear I was unable to help him but I felt his dynamic personality and got a very strong impression that if anyone on earth could get planning permission for a dwelling on that plot it would be Frank Taylor.

Then there was Nigel Bridge, a barrister with whom I worked several times. He had a habit of asking very long questions in re-examination. Re-examination occurs after cross-examination and affords your counsel the opportunity to try to repair some of the damage done to your evidence in cross-examination. Counsel is sometimes permitted to lead the witness a little – particularly if you are not appearing before a judge:

'Mr Smith, when you said Miss Jones was wearing a white hat, was the word "white" used in an approximate, imprecise sense, as a man might use – rather than a woman; I mean, was it used as a collective term embracing all the various shades of white, such as off-white, spilt milk, cream and the like, or did you mean absolutely "dead white".

'No, Sir. It was like what you said. Sort of white.'

'So, Mr Smith, what you are saying is that Miss Jones might have been wearing a cream hat?'

'Yeah. She might well have been.'

'I am much obliged, Mr Smith. I have no more questions. You may step down now.'

That's re-examination, but Nigel Bridge once put such a long question to me in re-examination that I forgot how he had originally couched it. I answered:

'I forget how you couched that question to me, Mr Bridge, but I entirely agree with the sentiments you expressed in it.'

That seemed to cause amusement all round.

The next time I was with Nigel Bridge was concerning a planning appeal in Derby. At the end of a long day, Nigel said to me just after the inquiry had finished:

'If we lose this one, Hilary, I will give up.'

We did, and he did – to become a High Court Judge.

I am told that in a case in which Gerald Eve was involved, a building surveyor on the other side was giving evidence and said: 'The wood was rotten'. Gerald leant forward and said to his counsel:

'Ask him what sort of rot it was. Which of these two was it?' He handed him a note. On it was written:

'*Coneophora cerebella* or *Serpula lachrymans*? It matters.'

When the surveyor had finished his evidence-in-chief, which had been given in a very confident manner, our counsel got up and asked:

'Mr Smith, you say the wood was rotten. Was the rot *Coneophora cerebella* or *Serpula lacrymans*.

'Was it what?'

Counsel repeated the question.

'I'm afraid I can't answer that.'

'So, Mr Smith, you do not know the difference between the two? How to tell one from the other?'

'No.'

He turned to the judge:

'My lord, the reason I am pursuing this is that the expert witness I will call, Mr Gerald Eve, will give evidence to show the importance in this case of distinguishing one type of rot from another.'

These Latin names only referred to Wet Rot and Dry Rot, but Gerald wasn't letting on. The witness's confident bearing in the witness box had been destroyed and, more importantly, his competence as an expert had been shown to be questionable.

Why was it important to tell dry rot from wet rot? Wet rot only attacks wet wood, but dry rot has water-carrying strands which can extend for some distance and suck all the moisture out of perfectly sound timber. The strands can even go through brickwork.

Chapter twenty-four

A Surfeit of Seifert

My first job concerning Central Area Redevelopment, as it was then called, was given me by my local council, the Walton and Weybridge Urban District Council. A development company had bought the vacant Nettlefold Studios, Walton-on-Thames, which used to make films. Their land was just behind Walton High Street and the company wanted to redevelop it partly as a shopping centre and partly as flats. The area connecting the studios to the High Street had been bought by the Council bit by bit over the years as they hoped to construct a relief road through the area at some future date.

The development company instructed an architect, Richard Seifert, to act for them. Colonel Seifert, as he liked to call him-

Richard Seifert, architect

self, was a wartime Lt. Colonel in the Indian Army – and probably a Temporary, Acting one at that. He was, at the time, the developer's favourite architect. The reason for this was that he was a genius at getting planning permission for the most unlikely things. For instance, he got permission for an incredibly tall block of offices, the Tolworth Tower (1962–6), at Surbiton on the Kingston Bypass. How did he get it? Bribery, I hear you suggest. Well, judge for yourself. Who was

Centre Point, 1962–6,
101 New Oxford St,
London, Richard
Seifert & Partners

the Local Planning Authority? Surrey County Council. To whom did he let the office block? Surrey County Council. Bribery? Call it what you like, but Seifert did provide a pretty good incentive for the County Council to grant themselves planning permission.

Then take Centre Point (1962–6), the famous, or infamous, office block which you see if you look eastwards along Oxford Street, London. Harry Higham's company built that but Seifert got the planning permission. How did he get planning permission? There was a traffic problem at that junction and Seifert promised the council that he would give them land for a big roundabout round the office block. Permission granted.

The office block failed to let for a long time and people said Higham kept it empty deliberately so that the rents would go up. I recall that Higham said he would sue people for libel if they accused him of such a thing. Such an accusation was absurd because developers like to let their offices even before they are built. The rental value will go up even faster at the rent review if they are tenanted, because the offices will be heated and kept in repair. The reason the block did not let, as is apparent if you see the floor plans, is that the building was built round a central core and the letting agents asked a normal price per square foot for gross floor areas. Once you starting working out the net *usable* space, – the building was very narrow – the deduction from gross areas was such an unusually large percentage that the rent per usable square foot increased substantially over market value. This, I suggest, did not say a lot for the architect's design.

Returning to Walton-on-Thames, my two clients, Hubbard, the Town Clerk and Bromley, the Borough Engineer and Surveyor, came with me to meet the developers and Brigadier Seifert – we promoted him regularly. Seifert was of slight build, fairly short, with a large, thin nose and dark brown eyes like currants, which constantly darted hither and thither so as not to miss anything anywhere. His hair was black and he wore thick, aggressive, horn-rimmed spectacles to match.

From the very start he took charge of the meeting in what I thought was a very domineering manner. It was not a discussion but a narration of what Major-General Seifert was going to do, when he was going to do it and how he was going to do it. When we started to query some of his proposals he got really shirty, ending up by saying:

'All right, then; if you don't accept my plan we will go it alone.'

One of the property company directors grasped his elbow and whispered to him. Although we couldn't quite hear all that he was saying we didn't need to – we knew:

'You must remember we haven't yet got planning permission and we don't own the land between the studios and the High Street.'

'Oh. Right. Well, gentlemen, the keyword must be partnership.'

Some partner!

My next meeting with Seifert was at his offices in London – just the two of us. It was the most unpleasant meeting I have ever had. He was charming, threatening and bullying in turn.

When I got back to the office, I went up to TSD's room to get it all off my chest. When he told me Seifert had been on the telephone to him, I burst into tears.

Seifert had telephoned him as soon as I left and said that I had been 'rude [not true], unhelpful, obstructive and uncooperative'. From Seifert's point of view, I suppose the last three epithets were true, in that I did not fall in with his plans or his values. I had had considerable experience in valuing retail shops and I was not happy with the positioning and linking of his new parade with the High Street. He also took it upon himself to usurp the valuer's role and proceeded to put forward patently absurd ideas as to what his land was worth compared with that of my clients. When I told him my own figures, which were substantially different to his, he got really angry.

I explained all this to TSD and he was very sympathetic and backed me one hundred per cent. He wrote an appropriate letter to Seifert, which he showed me. It was very clever and quietly supportive of me. I was so grateful, because I thought I was doing a really good job standing up to Lieutenant-General Seifert, making sure the shopping element was properly planned and sited, and that my clients got the full value of the land they were contributing to the scheme.

The next meeting with Seifert was at the development company's offices in Park Lane. It was going to be a crucial meeting where the broad details of the plan and the financial provisions were going to be agreed. I was determined that we would not be bullied by the general, nor my clients overawed, so I decided on a plan. I arranged for the meeting to be at 3pm and asked my two clients, Hubbard and Bromley, to lunch at the Bon Viveur in Shepherd's Market. We had a superb meal with wine, port and brandy. I reminded my two clients that *they* had the backing of their Council; *I* had the backing of my senior partner, and *we* had the land linking the studio land with the High Street. We held all the trump cards and had nothing to lose. Before we left, I bought each of us a corona-corona cigar – which is the longest you can buy – as I felt sure we would be offered one by General Seifert. We stuffed the cigars in our inside pockets and took a cab to Park Lane.

We were greeted and almost immediately offered a cigar – not very big ones, I was pleased to see. We all declined, reached for our corona-coronas and lit up together with some big puffs. I've never enjoyed a cigar more. Then we got down to work. Field

Marshal Seifert had got the message and we agreed all the plans and finance without any unpleasantness.

The Council got a multi-storey car park and a much-needed relief road. The developers got a retail store, a parade of shops linked to the High Street and several blocks of flats. It was then to my knowledge the only central area redevelopment scheme to make a profit.

When I looked on the back of my rate demand for my house (I lived locally), I was delighted to see each year at the end of a list of expenditure an item: *Credit; Walton town centre* with the annual profit from the scheme shown as a deduction from the rate in the pound.

It is amazing what a corona-corona can do for your confidence – and for your rates bill.

Chapter twenty-five

A Bloody Shambles

Shamble *noun* 1. A table or stall for the sale of meat (*Middle English*). 2. *Plural:* a slaughterhouse – *1542.* 'He was felled like an ox in the butcher's shambles' – DICKENS. 3. A scene of disorder – *1920s*

Another job I did for Walton and Weybridge Urban District Council concerned the central area of Hersham, just a mile or two from Walton on-Thames. I was given the job of acquiring a slaughterhouse. The owner-occupiers had put in a claim for compensation which, apart from claiming for the value of the freehold, included a large sum for what is known as 'disturbance', as the business was a going concern. Such a claim includes time spent looking for new premises, double rent for a period, removal costs, temporary or permanent loss of profits, loss of goodwill, loss of key staff if you have to move out of the district, and a mass of smaller items like renewing signs and stationery and sending out change of address cards.

The slaughterhouse was a pretty ancient one and I had heard that there were some sweeping new regulations to be brought in soon by the EEC. Although the slaughterhouse was a going concern, I wondered for how long it could keep going. To find out the answer to that question I contacted the Medical Officer of Health who provided an expert on the new EEC regulations. He walked round the slaughterhouse with me to say in what ways the premises would fall short.

I will spare you all the gory details but will tell you how the slaughtering procedure began. You may have had enough by then.

The animals walked in to the building one by one and each had applied to its head a gun which shot a bolt against it. This rendered the animal unconscious. The animal then had its throat slit and was put on a conveyor. Since the animal was still alive, the heart pumped the blood out of the animal thus giving people the white meat they expect pork to be. As you might expect, the place was literally a bloody shambles. I have never seen so much blood in my life.

I asked the inspector if this part of the premises would conform to EEC Regulations.

'You must be jocular,' he said. 'To start with, the floor, walls and ceiling must all have impervious surfaces. None of them has. Also, the ceiling is not high enough, the ventilation is almost non-existent, the facilities for flushing the floors with water are inadequate and the system for disposing of the water and waste is woefully insufficient and antiquated. The room is also much too cramped to install all these facilities. Furthermore, the men are not wearing the correct clothing, but that's not to do with the premises.'

'Could the premises be converted to conform to the regulations?'

'Not really,' he said. 'Apart from anything else, it's not big enough and it's not high enough. You'd need new larger buildings on a bigger site.

We went round the whole hotchpotch of buildings and it was the same story throughout. It was quite clear that when the EEC regulations came out, the slaughterhouse would have to cease operating. The site was simply not large enough on which to build a new abattoir, as they now preferred to call it.

The man acting for the abattoir owners was a solicitor, being a former Town Clerk from somewhere in Wales. I'll call him Mr Davis. I am not sure whether he was a director or merely an adviser, but although he had no qualifications as a valuer, he had calculated and submitted the claim, without any expert advice from a valuer.

I explained to him as gently as I could that I could not agree to any disturbance claim because of the imminence of the EEC Regulations. Although we would wish to acquire the freehold shortly, we could defer entering upon the land until the EEC Regulations came into force – by which time his company would have had to cease operations. Therefore we would not be buying a 'going' concern, because it would have already gone. Thus there was no disturbance claim because we would disturb nothing.

Instead of arguing the point, as a valuer would have done, Mr Davis became furious. Without dealing with any of the points mentioned, he called my attitude scandalous. There I was, acquiring a prosperous, flourishing business and not prepared to pay a penny piece for it. He stormed out of the meeting.

Mr Davis had not spent his life in Welsh politics for nothing. If you were up against a brick wall you did not beat your head against it. You sought a way round it.

The next time I appeared before the appropriate committee of the council, a councillor got up and said there was a delay in the acquisition of the properties which might hold up the whole scheme. In particular the agreement on the price to be paid for the acquisition of the slaughterhouse was being delayed because of a 'clash of personalities' between Mr Davis and Mr Eve. He suggested the matter be handed over to the Inland Revenue District Valuer.

I was quite flabbergasted and entirely unprepared for such a remark. I uttered something quite incoherent. The committee agreed to the councillor's suggestion.

I heard later that the amount of compensation had been agreed very quickly between Mr Davis and a valuer from the District Valuer's office. I checked with the Medical Officer of Health who confirmed that no one had consulted him. I was tempted to find out how much had been paid for disturbance and, if any had – as there must have been to settle the matter so quickly – kick up a fuss about it. On reflection, I thought there was no particular point in doing so, but it did occur to me that I might profit by going on a course to learn about Welsh political methods.

I might be an expert on compensation, but in the devious and dirty world of politics I was a complete ignoramus – and in this case I was left standing at the post.

Chapter twenty-six

A Brueghelian Feast

The drunken, bumbling characters who animate Brueghel's
lively scenes are the very opposite of the Italian ideal of
refined perfection. Yet it is Brueghel's work, based on real life
observation, that is the more real and human. Legend has it
that Brueghel would put on disguises in order to take part in
the peasants' rollicking gatherings.

The Art Book. Phaidon Press 1994

The great thing about attending RICS conferences was that the
expenses were allowable against tax. Thus, when we partners
were unfortunate enough (or, it could be said, fortunate
enough) to be paying eighty-three per cent tax on the top slice
of our income, sixpenny buns cost but a penny. So when
Mrs Thatcher came to power in 1979 and promptly reduced
the top rate of income tax to sixty per cent our penny buns went
up from seventeen pence a hundred to forty pence, an increase
of about two and a half times.

Thus, when, straight after Maggie's tax reduction, the Senior
Partner asked us partners who would like to go to the RICS
Conference at Jersey, almost nobody put their hands up. So the
RICS conferences I am going to tell you about now are all
pre-Maggie.

One such conference took place in Vienna and, on arrival,
Geoffrey Powell, a much travelled man, warned me that I would
be getting a lot of invitations to various functions, and if I did-
n't look out I would be stuffed full of delicious but fattening

Viennese pastries. He advised therefore that we accepted only those invitations that served wine.

Sure enough, when I went to register at the conference hall, within my bundle of papers handed to me at reception in a folder by some kind lady from the RICS, were about a dozen invitation cards. My wife and I scanned these and, strictly complying with Geoff's warning and in order to avoid gaining weight, refused any invitation that did not announce that wine would be served. We were delighted to see that out of a dozen invitations we had to refuse only one.

I gave the cards back to Geoff and after looking at them he said:

'So you didn't follow my advice then, Hillers.'

'Oh, yes, I did, Geoff,' I said. 'We scrutinised the invitations most carefully. Look.' I took back the cards and started to go through them one by one: 'Wien, wien, wien, wien…'

'You silly clot!' said Geoff. 'Wien doesn't mean "wine". It means Vienna.'

Later, we went to Budapest, Buda being on one side of the Danube and Pest on the other. The next day was May Day and the Communist Government put on a huge parade. It seemed to consist wholly of lorries and tractors and workers walking beside them all emblazoned with red and white banners. I have never seen so many people looking quite so grim and found it really depressing that a Communist regime could stop even schoolgirls laughing and giggling.

May Day Parade, Budapest, 1978

The most memorable RICS conference I went on was that centred on Brussels. We had a trip to Bruges where we took to small boats and went up the canals. Our fellow surveyors stationed on the bridges felt we looked a little hot – it was a warm summer's day – and were kind enough to shower us with cool water as we passed under each bridge.

The highlight of the conference was, fortunately, on the last evening, and was termed a Brueghelian Feast. It certainly lived up to its description at the head of this chapter. There seemed to be rather a long time set aside for pre-prandial drinks, so even before we took our seats for the feast most people were pretty well oiled. This was unfortunate for the minstrels who circulated round the tables with their bagpipes (as in Brueghel's picture) and received, from our long table at least, a volley of bread rolls. The minstrels were in no way nonplussed, such conduct no doubt being customary from 'drunken bumbling characters' at 'peasants' rollicking gatherings'.

Peasant Wedding Feast 1566 – Pieter Brueghel

However, Eric Strathon, my Senior Partner, seated at the top table due to his being a recent Past-President, felt most embarrassed and pretended that those young surveyors behaving so badly were nothing at all to do with Gerald Eve & Co.

Matters got worse when at the end of the meal someone passed me a note which said:

'This tablecloth is made of paper.'

I soon tore off a strip to make into a missile to launch against the bagpipers – after all, they are not everybody's cup of tea – and the rest of the table soon followed. Then the wife of the Director of The Civic Trust climbed up on to the table and asked me to dance. As I tried to explain to Eric Strathon afterwards, it would have been both ungentlemanly and discourteous of me to refuse.

Geoffrey Powell was a great one for practical jokes and he played them mostly on me. In the early days, I was liable to have a magnificent new Bentley waiting outside the office for me to test-drive, or have a salesman with Encyclopedia Britannica asking for me.

I remember once Geoff organising a pre-theatre dinner at a restaurant in the West End. There were eight of us; Geoff, his wife and two guests had a slightly earlier start to the theatre performance than we had. He went to pay the bill and, as he was leaving, said:

'I've paid the bill for all of us, Hilary. We can settle up later. Don't let them make you pay twice. The food's good here but they are really sharp. Given half a chance they'll try on anything to diddle you.'

'OK' I said, so when the waiter came up a few minutes later and presented me with the bill for the four of us I was more than ready for him.

'My friend paid for all of us just before he left a few minutes ago.' I said indignantly.

'I am very sorry, Sir,' said the waiter, 'but he just paid for the four of them.'

I was just about to get really angry with these reputed diddlers when I had a sudden vision. It was of Geoffrey sitting in the dress circle laughing his head off, thinking about my inevitable contretemps with the waiter. I paid up without a murmur.

It was almost at the very end of my professional career when I more than paid back Geoff for all the tricks he had played on me over the years, but the funny thing is that I started the whole process with no serious intention of playing a practical joke on him.

I was a member of the Savile Club and during their staff summer holidays they had a reciprocal arrangement with the Reform Club. I went and had lunch at the Reform and then went up to the library. There I saw on the writing desk some sheets of club writing paper and envelopes; so, on a sudden whim, I took one of each.

When I got back to the office I got my secretary to type a letter to Geoff on Reform Club writing paper. The gist of the text, which was headed STRICTLY PRIVATE AND CONFIDENTIAL, was that I was Secretary of the Reform Club and had had an approach from General Amin, President of Uganda, to buy the Club. The matter was obviously extremely hush-hush and Geoff should not let anyone else set eyes on this letter. His name had been put forward as the best man in England to advise on the value of the club and to carry out any negotiations which might be necessary.

I ended by asking him, for reasons of confidentiality, not to reply to this letter but to contact me at a certain telephone number. I signed myself FE Ingleby-Mackenzie because it was the first double-barrelled name that came to mind. Someone with that surname was a cricketer not too long ago and captained Hampshire. The telephone number I gave him was my own ex-directory direct line at the office, each partner having a telephone line to make his own calls. No incoming calls came on them. My telephone number was about three more than Geoff's so I assumed he would recognise the number immediately he read it.

It follows that when my direct line telephone rang the next day, I knew it was Geoff and I assumed he would say something like:

'Hullo, General Hilary Amin. How are you today?'

I therefore made only the slightest attempt to disguise my voice when I answered the phone with a somewhat suave:

'Ingleby-Mackenzie here.'

A conversation ensued during which Geoff indicated he was happy to act for the Reform Club. I couldn't believe he was being serious and it was all I could do to stop myself from bursting out laughing – particularly as my secretary, Alison Kirkness, nearby was giggling to herself quietly. It was left that I would get in touch with him again while I awaited further developments.

When I put the receiver down, Alison said:

'I think he has rumbled you and will turn the whole thing round to play an even better practical joke on you.'

'You may be right, Alison,' I said. 'I can't believe he didn't recognise my voice and that he didn't recognise my telephone number which is almost the same as his, though I must admit I would not recognise my number because I never dial it and never tell it to anyone.'

A week later, I went to the annual tennis tournament at Wimbledon, accompanied by a member. We went into the Members Room and there I espied some sheets of headed writing paper and envelopes. I stole one of each.

The heading on the writing paper was, I think, *The All-England Lawn Tennis and Croquet Club* and I dictated to Alison a letter somewhat on the following lines:

Strictly private and confidential

3 July 1978

Dear Powell,

I am, for my sins, on the Committee of the A–E and am writing to you from here for security reasons.

Thing are hotting up and moving very fast! 'You-know-who' is flying in to Heathrow on Friday this week in his own plane and wants a meeting with us at the Runnymede Hotel at 6 p.m. that evening. I explained to his aide that it was very short notice. He apologised but said he knew it would be impossible to change it as his boss was used to always getting his own way. I suppose dictators do.

Perhaps you would be kind enough to telephone me on the usual number and let me know if you are able to attend.

Yours sincerely

FE Ingleby-Mackenzie

I had suggested the time and place of the meeting because all the equity partners of Gerald Eve & Co. had arranged to spend the weekend closeted together to decide where the firm should be going in the next twenty years. We were to meet before dinner on Friday and had been booked in at – you've guessed it – the Runnymede Hotel.

Geoff telephoned me again and said he could attend the meeting. He said it was an amazing coincidence, because the funny thing was that all the partners of his firm had also arranged to meet at that very hotel that very same evening. I thought it really funny too, even funnier, and had great difficulty making my laughter inaudible.

Shortly before Geoff left on the Friday his son, John, came home for the weekend and Geoff told him about the job and where they were going to meet. John became very suspicious and said that it might be some sort of hoax. By the time Geoff arrived at the Hotel nearly all the other partners were already there. I had shown them all the correspondence and told them of the telephone calls so they were in on the whole hoax.

Geoff obviously felt very foolish – who wouldn't? – for being so gullible and, perhaps, annoyed with himself that he had been taken in by the chap on whom he had played so many practical joke over the years. However, if he ever lost his sense of humour, he soon recovered it, because he put copies of the correspondence on the firm's notice boards, one on each of our six floors of offices.

Looking back, I find it curious that the best practical joke I have ever played on anyone was not pre-planned and was entered into on the assumption that I would be rumbled instantly.

Chapter twenty-seven

Deus ex Machina

Deus ex machina (Lat.) The intervention of some unlikely
or providential event just in time to extricate one from
difficulties or to save a situation; especially as contrived in a
novel or a play. Literally it means "a god let down upon the
stage from a machine", the "machine" being part of the stage
equipment in an ancient Greek theatre.

Brewer's Dictionary of Phrase and Fable
– Cassells Publishers Ltd 1981

When Bill Haddock was about to retire, he said to me over a
drink one evening: 'Hilary, I think I should pass on to you what
I've learnt after spending twenty years interviewing people for
jobs in the firm.'

'Oh, yes,' I said hopefully, because I had a great respect for
Bill's qualities. He seemed to have no prejudices, was a pretty
well-balanced individual, and a tip, before he retired, about how
to select which people to take on would be really useful.

'Well, if you interview someone, you can tell whether or not
they are good at being interviewed. No more. No less. If he or
she seems to be a good sort, then take them on for six months
on probation. If after that time they are no good, tell them to
look around for another job. Don't feel bad about this. You are
doing them a favour. They are in the wrong job and the sooner
they get to the right job the better for them – and us.'

I am sure he was right because time and again we would ask
a chap to look around for another job and he would reappear in
a few years' time; not asking for his job back but as head of his

own development company asking for professional advice. Typically, he would have bought a piece of land, sold part of it for several times what he paid for it, and was having a little trouble getting planning permission for the last bit; or he would come to us to try to avoid or reduce a Betterment Levy on the profits he had made, or to get the rating assessment on his office block reduced.

The client I am going to tell you about now, Mr Greenland, did not seek an interview to join Gerald Eve & Co. If he had applied, he wouldn't have got as far as an interview because of his CV. He could hardly read or write. His hobby was buying 'chanceable' pieces of land and then sitting in his armchair writing out planning applications for various forms of development. In those days, there was no charge for submitting a planning application, and it may well be that Greenland's technique was to submit so many planning applications that the local planning authority would get fed up with him – and grant planning permission simply in order to stem the flow.

The way I got the job was this: Geoff Powell breezed into my office one morning and said to me:

'Hillers, old horse, would you like to go on a nice little jaunt into the country? Here's a lovely little job for you concerning a bit of land near Bath. I'd love to do it myself but I'm absolutely snowed under.' Geoffers always said he was absolutely snowed under. It was his favourite remark.

'Well, just this once then ' I said. 'You'll get me shot, you will.' A little jaunt down to Bath, my favourite city, sounded a lot of fun. He was out of the door in a flash muttering as he went: 'Absolutely snowed under…'

Little did I know that this case would involve weeks of work and my giving expert evidence before the Lands Tribunal. I should have known better. Our Geoffers, as a Welshman, was a pretty smart cookie and could smell trouble a mile away.

A few days later I drove down to a farm outside Bath and met Mr Greenland. He looked just like a typical farmer: strong, a big frame, a little overweight with white hair and a ruddy complexion. He had bright blue eyes which twinkled mischievously as he talked, with a broad 'Zummerzet' brogue; a man of few words but those he used were both apposite and witty, and many of his turns of phrase were unusual, even picturesque.

We drove off to look at a property he had bought in Lansdown which is an elevated district on the north-west outskirts of Bath and which estate agents could genuinely describe as a 'much sought-after area'. In fact, apart from Royal Crescent and The Circus, Lansdown is still today probably *the* most 'sought-after area'.

The property comprised a disused Victorian mansion in poor repair set in several acres of grounds. The island site was roughly oval in shape and was bounded all the way round by roads. There were lawns next to the mansion but all the way round the boundaries was a thick screen of trees of various species. The mansion was invisible from the road and was not overlooked.

The County Council had served a Compulsory Purchase Order on the whole property as they required it for a school. No objection had been made to the Order and I had been called in to advise on the compensation that should be payable.

When I first came back to Gerald Eve & Co. from Cardiff I did a lot of rating work but had done no compensation; so when such a case came along I sought some easy catch-phrase or watchword to sum up the whole thing. You will recall that because I was such a slow reader, and due to a 'slipped disc' was unable to carry heavy textbooks around, I confined my academic knowledge of rating to learning off by heart a sub-section of the Local Government Act 1948.

In compensation, I found it was the same as rating except for one important difference – you wanted more not less. And the watchword? I made it up myself: 'Compensation must compensate.' The rest was common sense.

In this case I found that Greenland had called in some nationally well-known planning consultants to advise him on how to develop the property – this was before the Compulsory Purchase Order was proposed. Unbelievably hamfistedly, the consultants had put in a planning application to demolish the mansion and to build narrow-fronted houses on plots extending all the way round the frontage of the site. Not only would such a development destroy all the trees and shrubs which surrounded the mansion, but the proposed development would be quite alien to, and totally out of context with, the nature of residential development in this high-class district. As you may have already guessed, the planning application was refused.

It was obvious to me that the appropriate development for this site, after demolishing the mansion, would have been to build on the site a large secluded block of flats in Bath stone to a very high standard of construction. The grounds, and in particular all the trees, would be left untouched. Once the building was up, a stranger passing by would not even know there was a block of flats there.

I went to negotiate with the Inland Revenue District Valuer, acting for the Council, who refused to concede that any development would have been permitted other than rebuilding of the mansion with a small extension. I don't think we disagreed about the value of the property on that basis.

So that was that. We had to take it or leave it. If we left it, that meant going to the Lands Tribunal – which could cost my client a lot of money if we lost the case. If we did lose, Greenland would be ordered to pay the other side's costs as well as his own, i.e. the fees for each surveyor preparing a proof and giving evidence, plus the legal costs of solicitors and counsel. As I slept on the problem and tossed and turned, I thought: thanks, Geoffers, for offering me a nice, relaxed little jaunt into the country. I decided to take the case to the Tribunal, so I

John Watson, Member of the Lands Tribunal

telephoned Greenland and told him so. He readily agreed.

The hearing of the case by the Lands Tribunal was in London at their premises in Hanover Square. Greenland came up for it. The member of the Tribunal who was hearing the case was Jack Watson, a Past-President of the Royal Institution of Chartered Surveyors whom I had met once or twice at RICS meetings. I remember nothing of the proceedings except that I was fortunate enough to have as counsel (my choice) Jack Ramsay Willis QC.

Some weeks later, we returned to Hanover Square to hear the decision being read. Jack Watson referred to the making of the Compulsory Purchase Order, the evidence given by the District Valuer and the refusal of permission for the planning consultants' application. Then he said that Mr Eve came down from London to solve all the planning problems 'like a *deus ex machina*'.

Jack Willis turned round to me and whispered:

'Look it up when you get back to the office, Hilary.'

The definition of this phrase is given at the head of this chapter. I gather that some Greek plays had extremely complicated plots and it was important even in those days for there to be a happy ending – like Hollywood films today if they are to do well at the box office. The easiest and often the only way to sort out all the complications and make everything end happily was for a god to descend from a 'machine', probably like Peter Pan does on a wire these days, and wave the Greek equivalent of a magic wand.

My magic wand seems to have been useful in this instance because Jack Watson added to my valuation of the mansion the sum of £5000 (about £50,000 in today's money, and many times this figure in today's land values) for 'hope value' i. e. the hypothetical prospect of getting the planning permission for the development I had suggested.

At the end of the decision, Jack Willis got up and asked for an order for payment of our costs to be awarded against the Council.

'Yes,' said Jack Watson, for he had awarded an amount of compensation which was much more than we had been offered.

Mr Greenland came up to counsel and me and shook our hands with an iron grip.

It had been quite a country jaunt.

Chapter twenty-eight

Full Marks for Marks

The wonderful thing about being a surveyor in a national firm was that you got to see all parts of the country. Indeed Gerald Eve sometimes used to start his proofs of evidence with:

'I have valued property in every county in England and Wales except one.'

If you asked him afterwards which county that was he would say:

'I don't know that there is one; but if, perchance, opposing counsel were to discover one in cross-examination, then that would be the one.'

The great thing about the claim, however, was that it made the judge or arbitrator wonder which county it was and thus brought home to him the unrivalled experience of the expert witness appearing before him.

Gerald also used to say:

'The two most efficient organisations in Britain are Joe Lyons and the Roman Catholic Church.'

During the time I practised, 1948–84, I would have altered Joe Lyons to Marks and Spencer. Not only were they highly efficient but they were a delight to work for.

I was introduced to M & S when TSD was asked to advise on some rating appeals to the Lands Tribunal that M & S had made from decisions of the Local Valuation Court.

The first property we went to look at was in Durham. If you've never been to Durham then you should do so. The *Lonely Planet Travel Survival Kit* calls it 'the most dramatic cathedral city in Britain' and the cathedral 'one of the greatest Christian buildings in the world'.

The most remarkable thing about the M & S store in Durham was that as you went through the front doors you came across a huge counter full of bananas. Why have bananas of all things at the very front? I was soon to learn the reason: If they did not have their banana counter in the front, the whole store would be clogged by banana-seekers. M & S had a policy of having identical prices for their products throughout all their stores in Britain. The fact that Durham was a very long way from Covent Garden, where the bananas came from, meant that their bananas were cheaper by far than any other shop in the district – or perhaps in the whole of the north of England.

That mystery solved, the rest was pretty straightforward. We went round the store, went to see the VO, got him to agree a reduction in the assessment, thus settling the appeal, had a look round the town and caught the train home.

Two other appeals were settled in a similar manner. One VO is reported to have said that negotiating with TSD was just one big compromise – at his figure. He certainly made some ingenious negotiating points. He was once negotiating the assessment of Swan & Edgar's (now a record store) which fronted on to Piccadilly Circus. The VO said that such a window display must be the most valuable in Britain.

'Oh no it's not, said TSD. 'It's virtually useless.'

'Pull the other one then. How can you possibly say that?' said the VO.

'Because all the world and his wife say "Meet you at Piccadilly Circus outside Swan and Edgar's". No shopper can see in the display windows at all as everyone is standing with his back to them.'

He has a point, you know.

The fourth appeal was at Felixstowe, Essex, probably the smallest town in which M & S had a store. The issue here concerned what we termed 'quantum' which was our jargon for 'quantity allowance'. Nearly all the M & S stores were freehold, except for those in new shopping centres, and thus the Zone A price applied to them was typically derived from rents passing on what we called 'standard unit' shops. The question was: should some discount be made to the Zone A price when applied to large shops and stores?

As you have probably guessed, we thought there should be a discount. You may say:

'Well, you would say that wouldn't you' as Mandy Rice Davies once said so memorably.

Perhaps, but we really did think that and I still do. In any case, we were paid to save our clients rates – within the rules of the game. One of my sayings is: 'The sign of a good rating surveyor is that he can wax indignant about the assessment of a property before he even knows how much it is.'

TSD decided that a good way to run the case would be to call in two distinguished rating surveyors each of whom was a Past-President of the RICS and the Rating Surveyors Association, each of whom had valued retail stores for rating, and each of whom would say that quantity allowances had and should be allowed when valuing retail stores for rating. One of these was Francis Fleury OBE and the other was John Postlethwaite. They were duly called as expert witnesses, followed by TSD who gave evidence about the particular store in Felixstowe.

Robin Laird, of whom more later, the member of the Lands Tribunal hearing the case found in our favour and awarded costs against the VO.

The last of the five appeals concerned M & S's Blackburn store. All I can recall about that appeal is that I wrote TSD's proof. He gave evidence – and again we won the case with costs again awarded against the VO.

It is well to remember that in each of these five appeals we decided when to settle and when to go to the Tribunal. There is a saying in advertising:

'Before you try to sell something make sure you think it's a good product.'

This certainly applies to rating appeals.

John White, the Surveyor of M & S, decided he could better spend his time buying property upon which to build new stores and enlarge existing ones, rather than dealing with the rating of all his stores; so at the next national rating revaluation in 1963 he handed over all the rating appeals to us, which comprised about 250 stores. My partner, Michael Hopper, took charge of that job.

Later on, John decided to give us the task of valuing all their assets. This massive job had previously been done jointly by two firms of agents, both of whom specialised in shops; so it was quite a feather in our cap – and an even bigger one when John said to us afterwards:

'It's the first time it's been done properly.'

The more I learnt about M & S, the more I admired Simon Marks, (1888–1964), later Lord Marks of Broughton. Whether or not he personally thought up all the innovations that were

Lord Marks of Broughton

made during his reign I do not know, but they were certainly revolutionary – and enormously profitable.

Who but a genius would have decided that customers could take back goods (not food, of course) they had bought and have their money refunded, no questions asked, simply because they had changed their mind? This meant that there was no need for fitting rooms and indeed there were none – they take up valuable sales space, need supervising and invite shoplifting. At a stroke, thousands of wives had no hesitation in buying, shirts, sweaters, ties, etc. for their husbands as Christmas or birthday presents in the certain knowledge that they could say truthfully:

'If you don't like it, dear, you can change it.'

Of course, in most cases the hubbies were polite enough, or lucky enough, to say:

'It's just what I wanted, dear.'

But, if they were not that polite or lucky, off they went to M & S to change the ghastly things for something decent – and, more often than not, buy a few other things there as well while they were about it.

Another change Simon Marks made was concerning the warehouse at the store. Previously, when an assistant was running short of some article, she (or he) would go up to the warehouse where the storeman would make her sign for the clothes she collected. Marks abolished all this so the girl went straight up to the warehouse and took what she needed. Just imagine all the bumf this saved for thousands of different items spread over 250 stores. People thought such a practice would lead to massive theft, but no such thing occurred and the savings in time and staff were enormous. This change was part of Marks's 'Operation Simplification' which did away with 26 million forms each year.

Under Marks's regime, they boasted that 99% of the goods on sale were British made; there was no such thing as a 'sale' – nothing was 'reduced'; and, being Jewish, he decided that as far as his stores were concerned there was no such thing as Christmas. He was also one of the first to ban smoking and if he saw someone smoking a cigarette in his store he would go and ask them to put it out.

Like all good leaders, Marks had his way of keeping morale high amongst the staff by giving the impression that he had the magic touch and always made the right decisions (which perhaps he did). This is borne out by two stories which I will now relate. The first is The Teddy Bear Story:

Marks put a limit of 250 on the number of stores. 'More than that and I would not be able to keep an eye on them all'. He had the habit of inspecting them rather like a general inspects his troops: that is to say, accompanied by a retinue and stopping occasionally to ask the odd question. One day he was carrying out his inspection when he stopped and looked at a teddy bear. He stared at it for a moment and then picked it up rather gingerly between finger and thumb, rather like you would pick up a dead rat. Well, not exactly like a dead rat, because it did not have a tail. On reflection, and I want to tell this story accurately, he probably picked it up by one of its ears.

He turned to his retinue and said: 'Is this our business?' The entire retinue took up their clipboards or took out their notebooks and wrote: *Teddy Bears – Is this our business?*

Marks replaced the teddy bear and moved on.

When the retinue got back to their various places of work, a massive enquiry was undertaken to find out if the teddy bear was making the profit required of it. After due diligence, they found out to their dismay that the unfortunate bear was not pulling its weight and should be what the Army would call 'dismissed the service'.

But how on earth, they asked each other, did Sir Simon Marks know that the wretched bear was not making the requisite profit? He must be a genius. He was, of course, but he was also a genius at convincing staff of his infallibility in leading the Company on the right road. Someone had been doing a little quiet research that indicated that teddy was less profitable than he should be – and Sir Simon had his own way of checking this out – and adding to his mystique.

I wish I could say that the bear lived happily ever after but I fear he was thrown out on his ear.

My second story is entitled: Hunt The Slipper:

An employee was sitting at his desk at Head Office in Baker Street talking on the telephone when a director walked in.

'All right, Jock, I'll see what I can do,' said the employee and rang off.

'Who was that on the phone?' asked the director.

'Oh, just the manager of one of the Scottish stores asking for some bedroom slippers. It's pretty cold up there and they're running short.'

'Well, perhaps you would be kind enough to get back to him and tell him to get the next train to London and report to my office.'

This was done and the manager reported to the director's office. The director first showed him a schedule showing the amount of slippers held in stock at every store in Britain as at close of business the previous (March) evening, the evening before that, a week before, and the same week as that last year. He then showed him the availability of slippers which had been ordered urgently from all their suppliers, together with their anticipated delivery dates. Next he showed him the weather forecasts for the various regions for the week ahead including the forecast temperatures.

'Now, Jock, armed with all the information that you now have, to which region would you send the limited stock of bedroom slippers that we can get in the next few days?'

'Well, Sir, it seems it's going to get a lot warmer in Scotland in the next few days and there's going to be a big thaw, but it's going to get much colder in the Yorkshire region; so I would send them there.'

'Well, Jock, I am sure you will be pleased to hear that that is exactly where we are sending them, so please never ring up to ask for more stock again.

'Now you can get the train back to Scotland, happy not only in the fact that we at Head Office here have some vague idea of how to run our business, but in the knowledge that when you get back to the comfort of your own home it will be much warmer there; so not only will you have no need to put on your own slippers but those few you have in your store will in all probability remain unsold until next winter.'

Chapter twenty-nine

Highways and Byways

Marks and Spencer, St Stephens Street, Norwich

My slow reading and inability to carry heavy textbooks did seem at times to have some advantages; the compensation claim for M & S on their Norwich store is a good example.

The store is on the corner of two shopping streets, and one of the frontages has a projection into St Stephens Street. The Council wanted to widen this street so as to make it into a dual carriageway and this necessitated acquiring the part of the store which projected into it.

John White of M & S gave me a plan of the store which showed all the counters and racks – the goods were mostly displayed on counters in those days. This plan showed the sales in pounds of the various goods on each counter for the past week. The counter frontage was shown in feet and thus you could work out the sales per foot of counter frontage.

For instance, if a Teddy Bear and a tie were the same price, and the teddy bear was six inches wide, you would want to sell twice as many bears as you would a tie which only took up, say, three inches of frontage – in order to obtain the same takings per foot of counter frontage. I calculated which counters were going to be lost by the demolition of the projected part and thus by what amount the takings would be reduced. Then I went to meet the District Valuer (DV).

We more or less agreed on the value of the property to be acquired, but when it came to permanent loss of profits my claim under this item was about ten times his. The whole claim was for about £250,000 which was several times as much as the DV had advised the Council that it would be.

The DV referred me to a book on compensation, which was far heavier than anything I could lug around with me, and in particular the case of *Somebody* v. *Thingumabob*. This case showed that the court had thrown out a claim for permanent loss of profits on the grounds of uncertainty.

I said I didn't know or care what the court said in that particular case regarding that particular occupier. We were dealing here with a store and with a firm whose profits had increased every year since the war. The continuing loss of profits due to the acquisition was not speculative or uncertain – the only thing uncertain was by how much they might increase year by year in the future. This was not just my opinion, it was what the market believed, as evidenced by the incredibly low yield that the shares had, due to anticipated increase in future profits.

The DV said he noted what I said and that he would report back to his client. The Council decided that in view of the cost of acquiring this property being several times what it had budgeted for, it could no longer proceed with the purchase.

So next time you are driving along St Stephens Street in Norwich and you see M & S projecting so far forward that there is no room for a dual carriageway (apart from a bus lane going the other way) you'll know why.

But don't blame me – I was only doing my job, mate.

The most unusual job that I did for M & S, and I suppose the most satisfying one – and the most fun – concerned their store in Oxford Street near Marble Arch which is, I think, in the best trading position in the UK.

Marks and Spencer, Oxford Street, Marble Arch

Shortly after the war, M & S applied for a building licence to extend the store. The civil servant at the Ministry of Works said:

'Sorry, but we only grant building licences to firms engaged in exporting goods.'

Back went the M & S chap, to Head Office, disheartened and empty-handed.

They put on their thinking caps and, a few days later, sent the chap back to the Ministry again.

The M & S chap said:

'Sorry to bother you again, but could you please tell me why you only grant building licences to firms engaged in exporting goods?'

The civil servant looked at him rather patronisingly and said:

'Because, my dear fellow, of the balance of payments crisis. We need all the foreign currency we can get.'

'I thought that might just possibly be the reason, so I have brought along with me the figures showing the amount of money we have taken in this store in the various foreign currencies during the last year. Here they are,' he said, tossing the piece of paper onto the large desk.

The civil servant picked it up and, as he took in the figures, his eyes widened in astonishment.

They got their building licence.

In spite of that success, on a site of that quality you simply cannot have enough space, so M & S were soon looking for a way of extending their shop further back. Unfortunately the rear of their shop was bounded by a highway, so the only way to extend further back was to stop up the highway at one end and build on it.

I was called in at the stage when M & S had negotiated with those entitled to object to the stopping up and had dealt with them by acquiring their interest – except one. The one remaining objector was a subsidiary company of Unilever whose directors had offices there. They were quite happy where they were and did not want to be bought out. They had to pay frequent visits to Unilever House which was on the North Bank of the Thames by Blackfriars Bridge, and they were worried that once the highway was blocked at one end (the east) the chauffeurs would take much longer to drive them there and back.

'There you are, Hilary' said John White, handing me the correspondence. 'That objection is the only thing standing in the way of a massive extension to our most profitable store in the country – a project which our chairman has set his heart on. See what you can do.'

I took a taxi from Baker Street back to the office thinking: why don't they give me any easy jobs? You did not need to be a genius to know the answer: because they can do them themselves.

Well, I thought, it's no good offering the objectors money; my clients have tried that. The only thing to do is to put oneself into the director's chair, so to speak, and see the whole thing from his point of view. What's he worried about? Taking longer to be driven to Head Office and back. By whom? His chauffeur. Right, I will go and see the chauffeurs.

Armed with a street map, I went to see the directors' chauffeurs and asked them which route they took to go to Head Office and back. They all went the same way and I marked it in red on the map. I returned to the office and, in the time-honoured Gerald Eve manner, called a meeting with several of my assistants and told them:

'Gentlemen, let us all air our ignorance,' and I outlined the problem.

What we had to do was to find a quicker route to get to Unilever House and back than that taken at present by highly experienced chauffeurs. The tricky thing was that if the highway was blocked up at the east end the chauffeurs would have to start their journey going due west – which was the opposite way from where they wanted to go.

'Easy' said one recently arrived graduate. 'If you want to know the quickest route, take a taxi.'

'That, Jeremy,' I said 'would normally be a very helpful and sensible suggestion – except that I have just done that on my way back from meeting the chauffeurs. Depressingly, the cabbie took the same route as the chauffeurs.' Sighs of despair all round.

However, *nil desperandum*, so we picked ourselves up and marked out three alternative routes. We decided we would test the routes in three cars; in each car would be one driver and one navigator with a stopwatch noting the times it took between the various junctions. In that way we might be able to take the best out of each route and make up one composite route. I had found out when the chauffeurs mostly made their journeys, so we had to ensure that we tested the routes at those times of the day.

Initial results were disappointing, but slowly, after two collisions and trying out some very unlikely detours, I was delighted that we found a route that was consistently quicker, if slightly longer, than that of the chauffeurs.

We prepared pretty plans showing in red the existing route and in blue the suggested route if the highway were closed. Attached was a schedule showing the average journey times there and back at different times of the day for the red and blue routes and the resultant savings in time.

I am not sure how pleased the chauffeurs were at being shown that they had not been taking the quickest route all these years, but the directors were very touched by how much trouble our clients had gone to deal with their problem and, moreover, to solve it. So they withdrew their objections to the stopping up of the highway. M & S were delighted.

So, today, when you wander into Marks at Marble Arch and delve deep into the store far from the madding crowd in Oxford Street, cheek by jowl with foreigners and their bulging wallets, remember that this extension alone probably solved the balance of payments crisis.

But would you also please spare just a thought for my two poor lads who each lost their No Claims Bonus on their car insurance in the course of serving their country?

Chapter thirty

Revenue Robin

'Ill fortune seldom comes alone.'

John Dryden, 1631–1700

I don't want you to get the impression that my surveying career was an unbroken run of successes, so it's about time I told you about two early cases that I lost. Of course, it wasn't my fault but you've probably already guessed I'd say that. I could blame counsel or the other expert witness but I don't. I simply blame the bloke on the Lands Tribunal who heard the cases, Revenue Robin, known by those who had not appeared before him as Mr Laird.

JR Laird, Member of the Lands Tribunal 1956–76

If there are any sportsmen amongst my readers, then they will be convinced of the truth of that statement by the following statistics concerning my appearances before the Lands Tribunal:

Played 7, Won 4, Lost 3.

Now comes the analysis of my results according to which member of the Tribunal heard the case:

	Played	Won	Lost	Success rate
Laird:	3	0	3	0%
Walmsley:	3	3	0	100%
Watson	1	1	0	100%

Need I say more?

My first appearance at the Tribunal was before the aforesaid Revenue Robin and was a rating appeal concerning Philips Character Shoes, being one of a parade of new shops recently built at the extreme end of Canterbury High Street.

At the Local Valuation Court I had criticised the VO's analysis of the rental evidence. He had added on to the rent the cost of the shop front by spreading the cost over the period of the lease. I had contended that in a position favoured by the 'multiple' traders, such as this, each trader had his own style of shopfront, and that if you considered the rental value of the shop 'in the open market vacant and to let', as you had to for rating, the value of the shopfront was negligible. My contention was that no other trader would add on any rental value for the shopfront, so the trader himself could get it by merely bidding an extra £1.

I told the court therefore that the VO had 'cooked up his valuation by making this wrongful addition.' The Chairman of the Court, who looked the type who wrote letters to the *Daily Telegraph* under the pseudonym 'Disgusted, Tunbridge Wells', at once took offence at my remark and went scarlet in the face. He said I was accusing the VO of 'cooking the books' and he was not prepared to have accusations of dishonesty made against the gentleman.

I was about say that that is exactly what I was accusing him of – dishonesty – but solely, and in particular, intellectual dishonesty; but then I felt that the nicety of my particularising the type of dishonesty of which I was accusing him would not be appreciated by Disgusted and would alas fall upon deaf ears.

So I apologised, said I had been misunderstood, and that in no way was I impugning the absolute integrity of the VO. But it was too late. The Chairman of the court had taken a dislike to me (and why not?) and confirmed the assessment. Hence the appeal to the Lands Tribunal.

At the Tribunal, I was determined to win my point about not adding on for the annual value of the shopfront so I made my plans accordingly:

I found out who was the chap at Philips Character Shoes responsible for the installation of their shopfronts; He was called an Inspector. I had learnt that if you wrote proofs of evidence for people not used to going into the witness box it worked OK for the examination-in-chief, but in cross-examination his real views came out, like the flood witness in Chapter 22. I was determined not to make the same mistake, so when I met the Inspector at the shop I resolved not to put words into his mouth.

I held my notebook at the ready and said:

'Are Philips Character Shoes shopfronts unique?'

'Oh yes; we have our own shopfitters who install the same distinctive style of shopfront in all of our fifty-five branches,' said the Inspector.

'Well, take this shopfront of Manfield Shoes next door. If you took that shop would you adapt the shopfront to your own use?'

'Oh no, we'd rip it out and put in our own.'

'Why is that? What's wrong with it?' I said. I was determined to make doubly sure that his evidence came from the heart and was not put into his mouth or suggested to him in any way.

'Well, Manfield's a different class of trade, isn't it? We sell the more inexpensive range of shoes, so we like to have more on display in the shopfront. So that stall board is too high and that means we would have to have a new plate glass window; and now you're talking big money. We'd also want to have more arcading so that it extended further into the shop, so the whole thing would have to be taken out.'

Game, set and match to the Inspector, I thought, writing in the last answer with delight and pocketing the notebook with a satisfied grin. It was splendid stuff – grist to the mill. I would be able to write a great proof of evidence for him, using his own words throughout.

But – you've guessed it – it didn't quite turn out like that on the day. *The best laid schemes o' mice and men Gang aft a-gley* wrote Mr Burns in 1785, and how right he was. They couldn't gang much a-gleyer that this one.

As the inspector entered the witness box he looked very nervous and I soon realised why. He was handed the Bible and asked to take the oath, but he whispered something to the court usher and handed it back. He was an Orthodox Jew. He took out a small scull cap and put it on the back on his head. After a short delay, another book was then produced and a different oath was said. The Inspector found all this rather embarrassing, which is not surprising.

Up gets our counsel, Patrick Browne, to take him through examination–in-chief. You are not allowed your proof of evidence in court in the witness box, so counsel holds it in his hand and takes the witness through it. You are not allowed to lead the witness. For instance, you certainly can't say:

'Lima is the capital of Peru, isn't it?' You can't even say:

'Is Lima the capital of Peru?' You have to say:

'What is the capital of Peru?' hoping your witness will give the right answer.

It doesn't always work out like that. I recall in *The Goon Show* an episode in court when the usher said:

'Call Sebastian Pericles Arbuthnott!' This was repeated half a dozen times, the voice getting fainter and fainter as the name echoed down the corridor:

'Call Sebastian Pericles Arbuthnott. Call Sebastian Pericles Arbuthnott. Call Sebastian Pericles Arbuthnott. Call Sebastian Pericles Arbuthnott.

Call Sebastian Pericles Arbuthnott. Call Sebastian Pericles Arbuthnott.'

Footsteps then echoed down the stone floor of the long corridor, first barely audible and then getting louder and louder
until you heard four loud plonks as he climbed the wooden
steps up into the witness box. There was a long pause and then
counsel said:

'What is your name?'

'Fred Smith.'

Patrick Browne says to the Inspector – we'll call him Levy:

'Mr Levy, have you had a look at the shopfront next door,
trading as Manfield?'

'Yes, I have.'

'If you moved in there would you use it?'

'Yes, I would display shoes in it.'

Counsel should never ask a question unless he knows the
answer. This counsel looked up at me. The witness was not giving the answer I had written down. Counsel in desperation
looked at the next paragraph in the proof.

'Mr Levy, what about the stall board? Is that suitable for your
own particular style?

'Well, I wouldn't put it in myself if we were starting from
scratch, but it's OK. I could live with it.' Counsel was talking to
a man who above all wanted to be amenable.

Counsel threw me another glance. If looks could kill I was a
dead man. He looked down at the last paragraph of the proof.
He staked everything on a final throw:

'Now, Mr Levy, we come to the arcading. Is *that* suitable for
Philips Character Shoes?'

Mr Levy, ever wanting to be positive and helpful, and to show
he was an adaptable sort of guy, said:

'Yes, Sir.'

'I'm much obliged, Mr Levy.' said counsel, slumping down
on his chair and giving me a final, withering glance

Revenue Robin, with a broad smile on his face, turned to
opposing counsel who half rose from his seat and said:

'No questions.'

We weren't surprised.

I was called next. My bit about the shopfront not having any market value didn't go down too well, but I produced rental evidence that showed clearly that the other side of the street was more valuable, yet the VO had put the same zone A price on each.

Our counsel also cleverly got the VO to admit in cross-examination that the assessments on the other side of the street were correct. After that admission, I thought we were home and dry; the basis for our parade of shops must come down because we were less valuable.

It was not to be so. Revenue Robin lived up to his name, dismissed our appeal and confirmed the assessment. We were all amazed. We had a con with counsel and I suggested the decision was 'wrong on the face of it'. In the body of the decision, Laird had agreed that the other side of the street was more valuable than ours, the VO had said the assessments opposite were correct, so, *ipso facto*, as the lawyers say, ours must be too high.

Counsel, who I thought was a bit of an old woman and a defeatist, wasn't having any of it. No, he could not advise going to the Court of Appeal (to which he was appointed some years later).

It was bad enough losing the appeal, but what was more annoying was that in reviewing the evidence Laird had omitted a large body of my evidence which had formed the basis of my case. Other members of the Tribunal would say, typically, that Mr X contended so-and-so because of a, b and c, whereas Mr Y contended the contrary because of d, e, and f. He would go on to say that he preferred the evidence of Mr X because...

Laird did not do that. He tended to select the evidence that supported his decision and omit that which did not. He did not carry out one of the four duties of a judge which is, according to Sir Francis Bacon 'to recapitulate, select and collate the material points of that which hath been said'. So I was hardly surprised when a fellow rating surveyor came up to me a week or two later and said:

'I read in *The Estates Gazette* Revenue Robin's decision in your Canterbury case and frankly, old boy, I am surprised you took it; it seemed to me you didn't have much of a case, if I may say so.'

Reading our Robin's decision, it certainly didn't.

As Mr Cooper of Rawlence and Squarey said so often:

'These things come in threes,' and come in threes they did as far as my appearing before Revenue Robin was concerned.

My next appearance concerned the modification or discharge of a restrictive covenant under Section 84 of the Law of Property Act 1925. Let me explain: On some housing developments the developer inserts restrictive covenants when selling the freeholds. For instance, on the estate where I live, built in 1929–30, you are not allowed to hang out the washing. On the estate with which I had to deal, my client was not allowed to build more than one house on his one acre plot; he wanted the covenant modified so that he could build a second house. To do this, you had to convince the Tribunal that the covenant had become obsolete.

When I had given evidence, including being cross-examined, Laird asked me, rather aggressively I thought, when I had first inspected the premises. I told him. Then two opposing witnesses gave evidence. I waited patiently to see if Laird asked either of them the same question. This would be the acid test, I thought. In the event, he did not ask them. I felt this was proof that he was biased against me and was not surprised when he found against my client and refused to modify the restrictive covenant.

I was so convinced of this bias that there and then I firmly resolved never to appear before Laird again. If I heard he was going to take a case on compensation or rating, I would settle the matter out of court.

Bishop's Stores was one of my rating clients; they were a family firm, and a delightful family too, and had a number of large grocery stores in the Home Counties and the outskirts of London. They had extended one of their stores at Ruislip and been served with a proposal by the VO to increase the assessment to what I thought was a sum substantially more than warranted. I was unable to negotiate with the VO what I thought was an acceptable assessment and had resorted to the Local Valuation Court. I forget what happened in the LVC but obviously the decision must have been unacceptable because, on my recommendation, an appeal was submitted to the Lands Tribunal.

As you can imagine, I waited with some trepidation to find out which Member had been allocated to hear the appeal. I was not too worried because the Tribunal had three lawyer members and six valuer members, in which case the odds were eight to one against having Revenue Robin. (In fact, they were not quite eight to one, as the lawyer members tended to take cases which involved points of law – which this didn't.) Thus I was both relieved and delighted when I heard in due course that the case had been set down for hearing by Herbert Prenzil Hobbs. I don't know why I recall his Christian names – because with a surname like that he was probably known as Jack. Later he paid Gerald Eve & Co. the compliment of sending his son there to get some professional experience. H.P. Hobbs was known as Halfway Hobbs because, pretty obviously, he had a reputation for giving decisions halfway between the two opposing values put forward at the appeal.

Jack Watson, whom you may recall in Chapter 27 called me a *deus ex machina*, had something to say about such decisions in his book, *The Incompleat Surveyor* (The Estates Gazette Ltd 1973).

> A common complaint about arbitrators has always been that
> they tend to 'split the difference' – i.e. add the two amounts
> together and halve the total. I understand that the same
> criticism is sometimes made of the Lands Tribunal. But… if

> the opposing expert witnesses are equally experienced,
> equally skilled and equally honest, and if there is no
> fundamental difference in the assumptions on which they
> have based their valuations, is it not possible – nay probable –
> that the answer is about midway between their opinions?

Personally, I'd go along with every word of that.

The day came for the hearing and we all gathered at the Tribunal's court at 3, Hanover Square. I had excellent counsel, an excellent case (my partners agreed with me on this) and was looking forward to giving evidence before Halfway Hobbs, a Past-President of the RICS and a member of the Tribunal who was highly regarded and well liked in the profession.

The court usher entered the room and said:

'All rise!'

We all stood – and in walked Revenue Robin. My heart sank. Counsel, solicitor, client and I exchanged glances and looked up to the heavens. We were going to need some help up there if we were going to win this case. However, we were taught in the Army that, when in a tricky situation like this, one must 'press on regardless'; so press on regardless we did.

The hearing seemed to go very well and, in spite of my previous experiences, none of us could believe that this plainly excessive assessment could be confirmed. Unreasonably, we were optimistic for, as Alexander Pope once said:

'Hope springs eternal from the human breast.'

Unfortunately the spring of hope proved not to be eternal. After Laird read out his decision – the spring dried out altogether. Assessment confirmed.

The issue before the Tribunal had been by how much should the assessment be increased to take account of the new extension and the question was: should a 'quantity allowance' be applied at the end, bearing in mind that the zone A price adopted had been derived from analysing the rents of much smaller shops.

Reading the decision, it was quite clear what Laird stated to be the decisive factor in coming to his decision:

'Mr Eve conceded in cross-examination that his clients increased the size of their shop so as to increase sales'.

What on earth has that to do with a quantity allowance? And *of course* my clients built more space to sell more goods. Why else would any shopkeeper on earth enlarge his shop? To make *less* sales? All my 'concession' indicated is that the assessment should be increased – which obviously I had done in my valuation. The question was by how much? – not whether.

After giving the matter due consideration, I decided that apart from any other fault he might possibly have, Laird lacked intelligence.

A while later, my partner, Michael Hopper, who dealt with the rating of C & A, Marks and Spencers and Debenhams department stores, had an appeal going to the Lands Tribunal. He told me Laird was down to hear the case.

'Michael, I advise you to settle out of court,' I said.

Michael later became senior partner of Gerald Eve & Co. and a very good one at that. On retiring from that position he became a distinguished member of the Lands Tribunal. You could not meet a fairer, more patient, kindly, better-balanced man than him. He refused to believe that any member of the Tribunal would not decide his appeal purely on its merits. He was also a brilliant expert witness and his judgment of cases later, when on the Tribunal, was impeccable.

He appeared before the Tribunal, came back and said the case went splendidly. Some weeks later, he came back to the office shattered. He had lost the case. He couldn't believe it and he said Laird's review of the evidence was not a balanced account of what occurred.

'I should have listened to you, Hilary,' he said.

But why should he? It is difficult to believe such things unless you have experienced them yourself. Perhaps Michael thought I might have irritated Laird in some way – Laird wouldn't be the first.

Over the years I have given the matter much thought and have concluded as follows:

1. Robin Laird was a good friend of my senior partner and 'Dutch uncle', Eric Strathon, and must therefore have been a nice chap.
2. He was not very bright and had a job sorting the wheat from the chaff when hearing a case.
3. He had an inferiority complex. This tallies in that he found in favour of M & S in the Felixstowe appeal. TSD, a top figure in the profession and two Past-Presidents gave evidence in that case, so Laird, by finding in their favour, was seeking the approval of the 'big boys wearing long trousers'. By contrast, slapping down younger partners of Gerald Eve & Co. and other top rating firms made him feel superior to them.
4. All this was done subconsciously. There is a saying that you only see what you want to see or what you are looking for. To test this, next time you go somewhere where there is a big crowd, think of a colour and look around. If you think red, everyone seems to be wearing red. Think blue and they're all wearing blue. It's no good trying this at a football match between, say, Arsenal and Chelsea because most of the spectators will be wearing their team's shirts, but try it at Twickenham at the Middlesex Sevens and you'll find it works.
5. He was not malicious, for in the case of *Horrocks v Lowe* (1974) it was held that 'unreasoning and blind prejudice will not constitute malice' if the person concerned 'honestly believed that what he said was true'.

I am sorry if you think I have been going on a bit too much about Mr R. Laird, but one of the reasons for writing memoirs is to get things off your chest – and I really feel a lot better after writing that last chapter.

And now for something completely different – a tale of unqualified and total victory before the Lands Tribunal. I can't wait to
tell you about it.

Chapter thirty-one

Arsenal 3, Leaseholders 0

The Royal Arsenal was founded in Woolwich in 1805, either because the Government was worried about Napoleon and decided to rearm, or perhaps to celebrate the Battle of Trafalgar in that year – after all this time I am not prepared to say which; but it wasn't until 1889 that the Royal Arsenal Co-operative Society ('RACS') purchased Abbey Wood, one and a half miles east of Woolwich.

Using their own building and works department the society put up 1425 houses on what became known as the Bostall Estate. 1285 of them were let on 99-year leases and when the Leasehold Reform Act 1967 was enacted, leaseholders were give the right to buy their freeholds on a basis laid down by statute. I won't bother you with the details.

When this Act was passed, someone on the estate decided to set up a Bostall Estate Leaseholders Association to look after the tenant's interests. The moving force behind this appeared to be a bearded gentleman whom we met, who had also been the moving force behind the formation of the Anti-Concorde Society. I didn't take to him.

I was very surprised that such an Association had been formed, because the RACS, as you might expect from the Co-operative movement, was a most progressive landlord; the society bent over backwards to keep a fair balance between the tenants' interests and its responsibility as landlord, being after all a non-profit-making organisation – or at any rate one which ploughs its profits back to its members. For instance, when the dispute about tenants' purchase prices came before the Lands Tribunal, the RACS made it plain that win or lose, they would pay all their own costs and part of the tenants'; a most generous gesture, which if it had not been made, would probably have avoided the dispute coming before the Tribunal at all – and saved them some thousands of pounds.

RC Walmsley, member of the Lands Tribunal 1960–81

To reinforce what I have just said, I will quote from the Tribunal's decision (Mr RC Walmsley FRICS) – not the whole of it, as it is some 11,000 words and 28 pages:

> Mr Eve told me that his instructions from the society in 1968 had been to prepare in respect of each house on the Bostall Estate a valuation under the terms of the 1967 Act... He had been told to exclude from his valuations anything in respect of a tenant's possible anxiety to purchase, and he had been asked that his advised prices should be fair as between the landlord and each leaseholder and also fair as between the leaseholders themselves

Who can say fairer than that?

In spite of the society's 'co-operative' attitude to the tenants, or perhaps because of their generous offer on costs, the Association, wholly unnecessarily in my view, took the dispute to the Tribunal. Three representative houses were selected as test cases, the first house, 14 Bostall Lane, being occupied by a Mr Siggs, so the case was called *Siggs and Others v. Royal Arsenal Co-operative Society Ltd.*

I collected together five surveyors in my department and we set off for the Bostall Estate. Our first and most important step was to decide the value of every plot let out on long lease (all

1285 of them) and thus subject to leaseholder's enfranchisement. Strictly speaking, we did not have to decide on the value of the site, but 'the site-element of the entirety value of each house'. Have I lost you? I hope not. Don't lose heart – it does get simpler.

There were three plots on the estate which had been sold, so we went to look at those and made a careful note of their size and position. We then drove round the estate, had a general shufti and repaired to the local pub where we took a drink and a private room.

Looking back on it today, I realise that it was at this moment that I laid down the foundation stone for our ultimate complete success in this case – by adopting the standard technique of Gerald Eve of dealing with a tricky problem: call a meeting and say: 'Now, gentlemen, let us air our ignorance'.

One of my helpers, Tony Taylor, had been in the firm for years and had an IQ of about 200. That was a good start. The rest had little or no experience (all the better) and were graduates.

I kicked off in the time-honoured manner by saying that it would be much easier if we put all the plots at the same value. Then we could have a drink and a meal and all go home. Was there any objection to that – not, I made clear, to the drink and the meal, but to giving every plot the same value?

AD (Tony) Taylor

Someone said 'yes' because some plots were more valuable than others. Why, I asked? Because they were bigger. Any other reasons? Because some were in more pleasant roads than others. Why were the roads more pleasant? They were tree-lined. Any other reasons? They were opposite a park. They had less traffic in them. And so on.

I made a note of each answer:

'Coming back to the size of the plots, how are they bigger?' I said.

Well, some were deeper than others and some were wider.

'Is a plot with sixty feet frontage worth twice one with thirty feet in this part of the world?' No, not quite.

'What about the depth of a plot? Is one a hundred feet deep worth twice one of fifty feet?' Certainly not; it's worth hardly any more at all.

We got down to work – and what did we produce? Let Mr Walmsley tell the story in his decision given on 17th September 1971

> In due course and after considerable thought Mr Eve had evolved a particular valuation approach based on capital site values in terms of £s per foot frontage ("pff"). He had adopted as the starting point in his calculations, the assumption of a plot having a standard depth and also having the deemed amenity characteristics of a particular estate road called Bostall Lane. His adopted site value pff for such a plot (its "formula value") was made to vary according to its frontage, ranging from £102 for a house with a narrow frontage of 15 ft. down to £72 pff for a house with a wide frontage of 28 ft...
>
> The actual formula values that Mr. Eve adopted are shown in the following table:...
>
> [These were given for twenty-three different frontages]
>
> ...Mr. Eve's next step was to make such adjustments, if any, to his adopted formula site value for the particular plot under consideration as he considered were necessary to bring into account its differing advantages and disadvantages compared with his standard plot, particularly as regards traffic, amenity and plot-depth, thus:...

Obviously, when we had worked out all the pluses and minuses for the various factors, we had to apply them to the three plots which had actually been sold in the open market to see if their 'formula site values' coincided with their sale prices. In the event, they coincided pretty well – and after a few minor adjustments they coincided even better.

I will skip the next 8,000 words.

> ...In the outcome I accept Mr Eve's figure of purchase price as calculated for 14 Bostall Lane of £308.

The next representative house was 71 Federation Road:

> ...The issue remaining is that of site value, and on this the valuers are not far apart, Mr Dennis [the Association's valuer] being at £1600 and Mr. Eve at £1640. In preferring Mr. Eve's figure I do so not only because I prefer his concept of "site value" but also because I attach weight to his greater experience in the ascertainment of site value for the purpose of the 1967 Act: Mr. Eve told me that he or his firm were advisers in respect of many ground rent estates, including Letchworth Garden City ("several thousand" rents), Western Ground Rents (about 10,000 rents), Hampstead Garden City Trust (over 4000 rents) and Gwalia (3,400 rents) as well as other smaller estates. I accept Mr. Eve's valuation...

The last of the three representative houses was 135 Rochdale Road.

> ...Mr Dennis disagreed with the making of adjustments to bring into account the particular advantages and disadvantages of a plot. "There is such a scarcity of plots," said Mr Dennis, "that the market would not give percentages." Mr. Eve's comment on this comment was that "the market higgles *all* the factors." I agree... *some* scheme of adjustments such as Mr. Eve has evolved is essential...
>
> I prefer Mr. Eve's figure for the same reasons as were explained in respect of 71 Federation Road and determine the price to be paid by the tenant of 135 Rochdale Road for the acquisition of the freehold interest to be £379.

★★★

For the assistance of the parties I would add a general comment. The way in which Mr. Eve has gone about his task of complying with his instructions to prepare valuations which are to be fair as between the landlords and each leaseholder, and also as between leaseholders themselves, shows immense care and consideration. Whilst my determinations of purchase price are, by their nature findings applicable to the three houses selected as being representative, nothing that was said during the six-day hearing has caused me to doubt the dependability of Mr. Eve's valuation approach generally.

As to costs, no order is being sought and none will be made. I was told that the parties had made an agreement which incorporated an arrangement for part of the tenants' costs being met by the society.

Mr Ashley Bramall (instructed by Messrs. J.D. Langton & Passmore, of Chislehurst) appeared for the tenants; Mr. Keith Goodfellow, Q.C. and Mr. Nigel T. Hague (instructed by Messrs. Joynson Hicks & Co., of Croydon) for the landlords.

I found the commendation at the end quite flabbergasting – especially after suffering all the pummelling from Revenue Robin in previous decisions and, although all my helpers in the firm were delighted with the result of their labours, I felt they should take most of the credit. The truth is that it was the initial 'airing of our ignorance' that did the trick. I can't take the credit for that, for that was Gerald's method – a method that has been passed down in Gerald Eve & Co. to this day. I suppose I can give myself the credit for remembering to employ it.

Every figure and every adjustment to our 'formula site value' had been questioned and argued out between us. Every stone of the foundation of our valuations had been put into place only after much thought and debate. The result was that in cross-examination, I could tell the Tribunal exactly why we had taken every step; every step was found by Mr Walmsley to be reasonable, simply because it had been reasoned out.

One thing I found rather disturbing was that once the decision was publicised, a number of people contacted me as they felt I was *the* expert on leasehold reform. The fact is that there was a lot of complicated law in it that my excellent counsel and good friend, Keith Goodfellow dealt with. On reading the whole decision again, just before writing this chapter, I was quite unable to follow all the legal argument. This may be partly due to my great age (77 now, 43 then) but I seem to recall that it was rather beyond me at the time.

Fortunately, Keith and his junior, Nigel Hague, were both excellent counsel. I had worked with Keith before on planning appeals, and his son and mine were both at the same prep school and played in the 1st XI football team. I suggested to Keith he should come and watch the matches as they were a cracking good side. He joined me, and I like to think it was our combined, vociferous encouragement on the touchline during the rest of the term that contributed to the team ending their season unbeaten.

Keith's favourite story was about the shopkeeper who received the lease of a shop that he was proposing to rent. He went through the lease until he came to the rent. He said

'I don't like this passage' and passed it to his solicitor. It read:

'The rent payable is ten thousand pounds per anum'.

The shopkeeper said:

'I have heard of paying through the nose, but this is ridiculous!'

★★★

Not long after that case, I was involved in another Lands Tribunal hearing involving Mr Walmsley. It concerned a compensation claim arising out of a Compulsory Purchase Order on land for highway purposes. Counsel for the solicitors acting for the claimant had recommended that I be brought in to give evidence.

The land was in a city up north, and the surveyor acting for the claimants was an estate agent in a large, well-established local firm. On meeting him (I'll call him Brown) I found that he had no experience in writing proofs of evidence or of appearing in court. Nonetheless, he was a very experienced estate agent, knew the locality intimately and had let many offices in the city. This last fact was important because our case was that, but for the proposed highway, the site could have been used for building an office block.

I decided that if I gave evidence rather than him, I would be vulnerable to cross-examination as I had never been to the city before. I thought therefore that the way I could help my client most would be to write Brown's proof for him and stay out of the box.

Mr Brown, like all the best estate agents, was intuitive and had an instinct for value rather than an analytical mind. I found that he had extensive evidence of office values but they were all in his head; he had no way of proving them in court. In no case could he produce the relevant lease.

After pondering for a while, I hit upon the solution: I would make a schedule and analysis of all the information on office rents in Mr Brown's head and send it to the District Valuer for him to agree or disagree prior to the proceedings. This I did and held my breath. The District Valuer agreed the entire schedule as being factually correct. I then wrote the proof with Mr Brown in attendance to keep me on the rails and to feed me all the information.

I attended the hearing and sat just behind counsel so that I was within whispering distance. Mr Brown entered the box and gave evidence but his manner was not too impressive. Indeed at one time Mr Walmsley, a very tolerant, courteous man, felt constrained to point out that he was irritated by one of Brown's mannerisms and would he be kind enough to desist from it.

Mannerism or no mannerism, Walmsley was too good a member of the Tribunal to let such a thing affect his judgment.

Brown knew his values and, with the District Valuer agreeing my schedule of office transactions, we were home and dry.

A couple of months later, Walmsley gave his decision. The District Valuer's valuation of compensation for compulsory purchase of the land was insufficient and would be increased to a higher figure. Consequently, costs would be awarded against the acquiring authority.

It was not a new experience for me to write a proof for someone else – I had written plenty for TSD – but it made me realise how TSD made the most of what I had written and what a superb witness he was.

Chapter thirty-two

'You're Always on Parade'

'When you're on parade, you're on parade. When you're off parade, you're off parade and you do as you like, but you're on parade now. Wake up!'

Regimental Sergeant Major Brittain, Coldstream Guards

One day I was referencing the ten-storey Head Office of the Times Furnishing Co in Holborn. I had got as far as the Director's Suite on the top floor when a man came up to me rather aggressively and asked me what I was doing there. I explained that I was from Gerald Eve & Co. and was inspecting the premises with a view … He cut me short.

'Gerald Eve, my boy. Wonderful firm. Did a marvellous job with the Jewish Bible College. Carry on, my boy. Help yourself.'

When I got back to the office, I went up to the Conference Room for lunch, where whatever partners were in lunched each day, and asked who had dealt with the Jewish Bible College. It transpired that one of my partners had dealt with the rating appeal on it. It was only a small assessment but, as far as I can recall, he had contended that it was occupied primarily for religious purposes and that that had been accepted. Consequently it was exempted from rates altogether. Thus, not only did the College not have to pay rates any more but all the rates they had paid for several years were refunded.

It was only a small assessment and barely profitable to deal with, but the Governors or Trustees were probably numbered amongst the top businessmen in the country. What finer advertisement could there be for the firm? It just shows that whatever job you do for any client anywhere, your reputation is always at stake. As my Senior Partner, Jack Powell, an ex-Army man, would say:

'Tip from your father, Hilary. You're always on parade.'

A very good example of this concerned a firm of chartered accountants whom Gerald Eve & Co. used to employ. They kept us happy for many years doing a grand job. Then I retired and gave them my tax return to do which was very simple but they made a complete mess of it. Accountants being very cost-conscious, they had decided to put some young lad on my little job. That is fair enough, but if you do that it is essential that a partner check his work. That is why at Gerald Eve & Co. a partner signed every letter an assistant wrote – even if it was just an acknowledgment.

The young lad given my job made a complete hash of it, so I looked into the whole thing myself – it was not difficult – and got the tax calculation right. I sent it back and he agreed it was now correct, but there was no apology. In came the fee and I wrote to the senior partner and told him to stuff it. I wasn't quite as blunt as that: I sent him the papers and asked him if he was satisfied that the quality of workmanship came up to his firm's standards. I pointed out that I had had to start the job again, do the whole of the work myself, and that it would have taken me far less time to do it if I had never instructed them. So perhaps *I* should charge *them*. He apologised and said in the circumstances they would not make a charge.

A while later, I happened to tell one of my former partners about it and, shortly after, I noticed that the firm had changed its accountants. The firm had lost a good client. If you don't have 'quality control' the reputation of your firm is as good as your worst employee on his worst day.

Whilst on the subject of fees, I remember saying to John White of Marks and Spencer that I was sorry the fee had come to so much money on a particular job – the one in Chapter 29 where we got the highway stopped up. John said the firm's policy was this:

'As long as we feel you are giving us good value for money we'll stay with you. The day we think otherwise we will go over the road.'

Who could, or should, ask for more than that?

Some people say that large firms are bad payers but I never found it so. Take M & S: I once mentioned casually to John that their solicitors had not paid our bill for one of the Lands Tribunal appeals. He got onto them to see if there was any good reason for it. There wasn't, so he sacked them – even though they were a large, well-known London firm. M & S always paid our bills by return of post.

At the other extreme was a firm who made earth-moving equipment. I remember the firm's boss saying to me that his second-hand earthmovers were the cheapest on the market – till the customer wanted some spares. He'd bought them all up and made his unfortunate customers pay through the nose for them.

He had not paid our bill even though he had had it for months. Fortunately, his company secretary was helpful. He said on the telephone to me:

'Mr Eve, would you do me a favour? Send me a solicitor's letter.'

'Why is that? I said. 'Is that really necessary?'

'Yes it is, because I have strict instructions only to pay when I receive a solicitor's letter.'

The letter was sent and the cheque came by return of post.

One of our clients, Cyril Lord Carpets, very quickly became a very big name in carpets and I dealt with their rating assessments in Oxford Street W1 and elsewhere. I was talking to the company secretary on the telephone one day when he said to me:

'Mr Eve, your firm have done a splendid job and saved us a fortune in rates and I would very much like to pay your bill. Could you get it round to me by hand today? I won't be able to pay it next week as, to put it bluntly, we're going bust.'

I did – and they did.

The partners or Gerald Eve & Co., probably like many professional firms, were more interested in the job in hand than in sending out accounts. It was a real struggle to get people to send out accounts – and the trouble was that it was the partners who were at the root of the trouble. One partner, whom I am sure will be glad to be nameless, handed over to me a client's file for me to take over the work. Although we had done many jobs for the client, no accounts had been sent out for six years. I decided to do something about it.

First I designed proforma accounts for all the most common types of jobs, such as valuations, rating appeals, rent reviews, planning appeals. These were put in a small metal chest of drawers, duly labelled, one on each of our six floors. All assistants were told it was *their* duty, not the partners, to fill out the proformas when each job was finished. The completed proformas were to be put in the partners' letter-books for them to amend or sign.

This arrangement certainly worked well with the partners and associates in my department and I feel certain it made an improvement elsewhere in the firm.

I do like simple solutions to difficult problems and was intrigued by one solution told to me by a chartered accountant who was sunbathing in the deck chair next to me on holiday in Sicily.

He worked for an American company who made machines which were manufactured in 125 different countries, and he was a member of the trouble-shooting team who was sent to factories not performing properly.

On one occasion, the team was sent to their factory in Turkey. They were welcomed at the airport by the local managers and taken to a private room in a hotel where they were given a big, boozy lunch. Then the girls were brought on.

To the hosts' dismay, my friend waved the girls away and said it was now time to visit the factory and deal with what had caused their visit – the frequently late deliveries of their machines to their customers. The Turkish managers were quite bewildered; this was an entirely new way of conducting business – and not a very pleasurable one at that.

At the factory, it was explained that the late delivery of parts was not their fault at all – it was the fault of the suppliers of the various parts, who were often late in supplying the part of which they had run out. They could, of course, have ordered large numbers of every part but that would have meant having vast stocks of everything, which would be expensive.

The remedy was magical. Let's deal with one part – a 3" bolt, for instance:

Q. How many 3" bolts do you use a week? A. 100.

Q. How long is the period from order to delivery? A. One to two weeks.

Q. In what numbers are the bolts delivered in? A. Cartons of fifty.

Right: order four cartons (two weeks supply) and when they arrive put them in a bag and tie it up with a fastening that includes a stamped postcard to the supplier ordering four more cartons. When the storeman is asked by the factory for more of the bolts he undoes the bag and puts the postcard in the post-box provided. The bolts will be delivered, you say, in a maximum of two weeks, so they will arrive by the time you have run out. Then do the same again. Adopt this system for all your parts.

Simple and brilliant. Can you imagine the savings that this procedure made in stock and in delays when adopted in 125 different countries?

I must have told this story to my senior partner, Bill Haddock, for he asked me to revise the firm's filing system. The present system which I think had been operating since 1930 was as follows:

When a new job came along, three index cards were made – one for a geographical index, one for a clients index for the accounts department, and one clients index for the 'dead' files in the basement, which contained the dead file number assigned to it on its death. I decided one of the client's cards was redundant because the dead file number could be put on the accounts department copy. Thus I was able at a stroke to throw away several thousand cards together with their cabinets.

My next task was undertaken at a time when computers cost a fortune and had to be put in vast air-conditioned rooms, so I decided on a strip system. When a new job came, each secretary would have been given a card showing the codes to be entered on the strip.

1. Select the appropriate coloured strip for Type of Job e. g. blue for Valuation, green for Planning, pink for Rating etc
2. Write Name of Client
3. Write Address of Property
4. Look up Type of Property Code. E.g. for a shop ('commercial') write COM 3
5. Look up code for Sub-class of Job. E.g. for Rent Review, write RR
6. Write File No. with partner's prefix. E.g. HE 1234
7. Write an identical duplicate strip.

The strips were then gathered together each week, slid into a metal holder, photocopied and sent to every partner and Associate as a New Jobs List.

After that the strips were taken to a central area in the office and put into the appropriate holder in two free-standing carousels, where you could turn the strip holders or metal pages and push the whole thing round as well – rather like you see postcards displayed in, or often outside, shops.

Thus if you wanted to look up, say, oil refineries you would look at the strips indexed by Type of Property. There under IND ['industrial'] 16 would be all the oil refineries we had ever dealt with –and if it were a Planning job you would look for the green strips. You could then ask for the file or see what the partner's prefix was on the file and contact him for advice.

In the second carousel was the Geographical Index. There if you had a property in say Shaftesbury Avenue, London, you would see all the properties in that street in order of their street numbers, whether they be, shops, offices, theatres etc.

Again, you could use the strip colour to pick out the Type of Job: say, pink for Rating. If you wanted to make a record of the whole page, you simply lifted and unhooked the metal page and copied it on the photocopier nearby.

It worked very well for many years and I expect after I left it was pretty easy to transfer the system onto a computer.

Another task that Bill gave me was to improve the firm's English. I was a bit worried about this because I suspected that the note I sent round to all partners and professional staff would be criticised for its bad English, grammar or punctuation. Like all notes, I felt that wherever possible it should not be wordier than that which could be contained on both sides of a page of A4. There was a tradition in the firm of keeping things short. This is what my partner Billy Oliver wrote in an article for GEN, our house magazine:

> When I presented my first report to Gerald it consisted of
> five closely written pages of foolscap. He took one look at it,
> tore it in half and threw it in the wpb. "Oliver" he said, "I
> abhor circumlocution; never use five words where two will do.
> Write your report on one sheet, double spacing, and I'll read
> it". When he had left the office I retrieved my effort and
> condensed it to the required length.

I got hold of Sir Ernest Gowers' little book, *Plain Words* which
was written for civil servants shortly after the last War, and
decided to confine my note to extracts from that book and from
the edition of Fowler's Modern English Usage revised by him
(Oxford University Press 1965) – so much more amusing than
the present (third) edition of 1996.

I remember Sir Ernest's commendation of the Egyptian Civil
Service whose motto some thousands of years ago was: 'Be
short, be simple, be human'. He quoted a letter written by an
Egyptian Minister of Finance to a senior civil servant:

> Appollonius to Zeno, greeting. You did right to send the
> chickpeas to Memphis. Farewell.

I have always thought that sloppy use of words means sloppy
thinking. There are always 'in' words which people use as a sub-
stitute for thought. At present it is difficult to read some company chairman's speech without coming across the words
'ongoing' 'focus', 'cutting-edge', 'key', 'strategy', 'moving forward'; and problems no longer have to be solved – they have to
be 'addressed'.

Some job advertisements are hysterically funny: take the
Appointments Section of this week's Sunday Times:

> 'The role will provide vision, strategy, control, and a
> governance perspective.'

A what? In any case, roles can't provide.
Here's another:

> 'You will need to be comfortable with ambiguity.'

Isn't that a bit ambiguous?

Try this one:

> 'You will have experience in addressing business objectives...
> Ability to transform strategic objectives into tactical targets.'

Note that business objectives are to be 'addressed' but strategic objectives are to be 'transformed'.

Finally, bearing in mind Gerald's maxim of never using five words when two will do, what would he do to this advertisement?

> 'This is a senior leadership development position with an
> emphasis on organisational change and leadership
> development... The focus of this position will be the roll out
> ...of the global leadership programme... The ideal candidate
> will currently be in a leadership development position... They
> [sic] will have gained experience in overseeing leadership
> development... This is an excellent strategically focussed
> leadership development role...'

You probably want to know what job they are offering:

> 'Head of Leadership Development.'

Planners have their own jargon. Everything had, and still has, to be 'identified'. Also, when considering a scheme of central area redevelopment, people would ask if it were 'viable', without bothering for a moment about what the word meant. Really, it means 'capable of independent existence', but people would use it because it was 'trendy', leaving you to work out whether it meant feasible, plausible, workable, profitable, politically acceptable, or something else – or perhaps nothing.

Today we have 'sustainable' development. What does that mean? A development that lasts for a reasonable period, as opposed to a sandcastle on a beach? And what about 'affordable' housing? Affordable to whom? Every house, mansion or palace is sold at an 'affordable' price to someone – otherwise it would not sell.

271

Let me leave the subject of plain words with a thought as to the serious consequences that can follow from sloppy wording. Fowler mentions the following:

> (From a notice in a public park):
> *Any person not putting litter in this basket will be liable to a fine of £5.*
> Those who have no litter must, it seems, go and find some.

Chapter thirty-three

Arbitrator at Leisure

For some reason which I have never discovered, my name was given to the RICS as an expert on 'Leisure Properties'. Perhaps it was our telephonist. Consequently I was appointed arbitrator by the President of the RICS on several leisure properties of which I had no previous experience – as a valuer. Luckily, I was not appointed as 'an independent expert' who can carry out his own valuation if he wishes without consulting either party; indeed sometimes no valuer is appointed by either side so he has to value using only his own expertise.

On the first occasion, the property was a casino on the Isle of Wight, where I had never been, and fortunately I was able to listen to the evidence of two expert witnesses before making my award.

Another time, I was asked to arbitrate on the value of part of the basement of Annabel's, the nightclub in Berkeley Square, London. I decided that, because of its limited access, if the basement had not been let to Annabel's its only other possible use was as storage, so I awarded a figure slightly above storage value, even though Annabel's clients were probably spending a fortune drinking down there. The landlord's surveyor later asked me, rather unkindly I thought, whether I had now been made a life member of Annabel's.

As an arbitrator or judge, you are lucky if you please half the people half the time.

Then there was a cinema in the West End of London used for television programmes broadcast before invited audiences.

I think it the duty of an arbitrator to try to work himself out of the job, if he can, by seeing if he can get the parties to agree, rather than go to the expense, as often happens, of both parties employing solicitors and counsel.

With this in mind, I called a preliminary meeting of the two valuers to find out the extent and cause of the disagreement. It turned out that the valuers could not agree on whether the audience seating belonged to the landlord or the tenant. It transpired that this issue was the only reason why the two valuers could not agree on the rent to be payable under the lease after the rent review. I suggested they went away and took legal advice. They did, and settled the matter without further recourse to me.

In one arbitration hearing, where both parties had instructed counsel, the landlord's counsel informed me that there was a point of law at issue in this case. He then explained in detail the point of law at issue which I found quite fascinating. I felt sure it was not beyond my intellectual abilities to pronounce upon this point of law when giving a discerning and distinguished judgment. Always interested in the law, I settled down happily to hear the submission that each counsel would make before me. I would then give a judgment which, because of its sheer intellectual brilliance, would be accepted by the solicitors and counsel on both sides. If not, the judgment would be confirmed by the superior courts as being both well conceived and wholly sound.

When the landlord's counsel had finished detailing this intriguing legal point, a preliminary point of law which needed to be settled before the valuers could give their expert evidence, I asked the tenant's counsel what he had to say about the matter. The landlord's counsel interrupted to say that he did not propose to argue the point before me. He was going to argue the point in a court of law because 'with the greatest respect' (and we all know what he means by that) he wished a judge to pronounce on the legal argument, rather than a mere (he didn't say 'mere' but I knew he meant it) valuer appointed by the President of the RIC S.

My face, which should have been as expressionless as the Sphinx, fell. I felt just like a kid whose ice cream had suddenly been taken away from him after the first lick.

I thought it would jolly well serve them right if they got a really rotten judgment.

My next arbitration concerned the rent review of a squash club. The landlord's valuer had once worked for Gerald Eve & Co. As I would expect, his schedules of comparable properties and valuations were all very well presented. I was impressed. Then up got the tenant of the squash club; he was scruffily dressed and his hair was tousled: it looked as if it had not had the attentions of a brush or comb since he had risen that morning.

'Sir', he said, 'Mr Smith's evidence may seem all very impressive' (he must be a mind-reader) 'but he doesn't know a thing about squash courts. The rents he is quoting are all paid for courts which are far superior to mine.'

He then went on to explain all the niceties of squash court construction, including insulation. He said the materials of which the court's walls and floor were constructed materially affected the bounce of the ball. On this count his court was inferior in every respect. He went further and said that really good squash players were not prepared to play in tournaments on his courts.

The landlord's valuer had no answer to this. Clearly this ex-Gerald Eve & Co. surveyor had not stayed long enough with the firm to learn that he should always try to know more than the opposition about the subject at issue.

The tenant then handed in a scruffy, hand-written valuation. Once I had deciphered it, I accepted it in its entirety.

Not all my valuations of 'Leisure Properties' were made as an arbitrator, for one of my clients was Paul Raymond, the

Paul Raymond

proprietor of Raymond's Revuebar, now closed.

His ex-wife, Fiona Richmond, used to write a monthly column for a girlie magazine – it might have been Penthouse – wherein she would describe in explicit and minute detail her sexual experiences when visiting some foreign city. It must have been somewhat embarrassing for her son who was at school with mine. Imagine a boy coming up to him and saying:

'Your mum didn't half have good time in Hamburg last month, Raymond. Do you know what she did? She...'

Once, whilst riding in his huge chauffeur-driven Rolls Royce, I happened to mention to Paul Raymond that his Revuebar must be doing pretty well. He said:

'You don't think I make all my money just from Raymond's Revuebar, do you?'

As he gave us more and more of his properties in Soho to deal with I began to realise the source of his wealth. One such property that he owned was let to a tenant who ran a sex cinema. I was asked to deal with the rent review which was going to arbitration. To do my job properly, not only did I have to inspect carefully my client's property but other sex cinemas which were rented and quoted as 'comparables' by each side.

I took two young lads with me to start this thoroughly boring and burdensome task and went to one of the sex cinemas cited as a 'comparable'.

'Lads, you measure up the foyer and ticket office while I go into the auditorium and count the seats,' I said.

I walked in through the doors and found a seat. I looked up at the screen which was about thirty feet wide; it was occupied from one end to the other by an erect male organ. I sat there fascinated to see what happened next... Plenty.

After about twenty minutes my two lads came in with their clipboards and asked me how many seats there were in the auditorium.

'Oh, the seats, ah, yes. The number of seats. I will count them right now,' I said.

That incident reminded me of the undergraduate at Cambridge who was watching a film – it might have been 'Extase' starring Hedy Lamarr. Miss Lamarr started to undress by a pool and, just as she was about to discard her last remaining bit of clothing, a train came along the railway line just in front, obscuring her. By the time the train had passed she was in the pool. The man sitting next to the undergraduate said:

'I must have watched this film twenty times and every damned time that train is bang on time.'

Another 'comparable' I had to look at was a peep show. The lad I took with me was a Cambridge graduate newly arrived at the firm that very morning. We went to the property in question which had a long, narrow arcade. On either side of the arcade were large photographs of unclothed ladies. I walked down the arcade and when I got to the end turned to my new lad to say something and found he wasn't there. I retraced my steps and found him still in the street; he was too frightened to go in. I took him firmly by the arm. I said:

'Come on, Jeremy. This is business not pleasure. Our job is to decide this property's rental value – and we can't do that without going inside.'

I marched him to the box office, showed my letter of authority and we went though the door. In front of us was a large walled circle with a passage going all the way round the circumference. A notice invited you to go to one of the openings in the wall which had small doors in front of them. You were told to put a sum of money in the slot and the door would open for a couple of minutes or so.

I put in my money and the door sprung open. I looked in the small window and jumped back in surprise as I saw a triangle of pubic hair wriggling just behind the window. Jeremy took a sidelong glance and backed away in alarm. The girl moved away and went into the centre of the ring. I looked through the window and two nude girls were dancing. In a remarkably short space of time the door clanged shut and Jeremy hurriedly almost ran out of the door, through the arcade and into the street.

I often imagine him arriving home that evening and his mother saying to him:

'Hullo, dear. How was your first day at the office? What did you do?'

Chapter thirty-four

A Pompous Bastard

One of the cities on which we were called in to advise on the redevelopment of their central area was Middlesbrough, on Teesside. Whenever I write the name of that city I am reminded of the publisher who published a book about the city. When all the books had been printed and bound, someone looked at the title on the front page. Middlesbrough was spelt 'Middlesborough'. All the books had to be pulped.

I think that was a bit tough on the copy-editor: first, because the spellcheck on my computer says it's Middlesborough, not Middlesbrough (there it goes again, underlining it in red): second, because the names of certain English boroughs nearly always either end in 'borough', like in Scarborough, or 'burgh' like in Aldeburgh. I say nearly always, because otherwise some clever dick will write to me via my publisher telling me of another place in England that ends in 'brough'. Well, don't bother. I concede there is another place (or places) that end in 'brough' but I just can't quite recall them at this very moment. (Don't mention Netherbrough or Brough to me because they are towns in Orkney and the inhabitants there would not like to be termed English.)

Mind you, the locals in Middlesbrough don't exactly help copy-editors to get the spelling right: if you go to watch a match at Middlesbrough Football Club you will hear their supporters shouting:

'Come on, Borough. Come on Borough!' None of them shouts:

'Come on Brough!'

The mishap we had at Middlesbrough was of a different nature. Our team comprised myself, Margaret Thomas who was a very bright geographer, and Dr Nathaniel Lichfield who had recently joined the firm as an Associate. The good doctor was a planning consultant who had written a book called, or about, 'Cost/benefit Analysis'. If you don't know what cost/benefit analysis is I am certainly not going to bother to explain it to you. I will, though, leave you with a little crumb – it is something to do with analysing the cost of something compared with its benefits. If you want to know any more, go buy the book.

Our team had spent some time up in Middlesbrough (there goes my spellcheck again) and this visit was to be the occasion when Dr Lichfield would outline his plan to the Council and its officers. Unfortunately, Nat, as we called him, missed the train from London, so Margaret and I had to decide on what plan would be presented to the Council. For some reason (did Nat make a habit if missing trains?), Margaret had lots of felt pens and paper with her on the train journey, so we got cracking and drew up a plan for central Middlesbrough. I presented it to the Council and all went well.

A few months later, I flew up to Middlesbrough via Teesside Airport. It was interesting to notice in those days that above every town there was a cloud – caused by the smoke from coal emerging from hundreds of chimneys.

I remember the Borough (or should I say Brough?) Engineer and Surveyor telling me that it was very difficult keeping white collar workers as they tended to find work in the south. I said the trouble was the air was so filthy due to the smoke from the chimneys of ICI at Billingham and also those of the steelworks. He pointed out that it would cost a fortune to put that right which would make those plants uneconomic when compared with their competitors on the Continent. They were the two principal employers in the area. Today, I believe the smoke emanating from these chimneys has either ceased or been purified.

On my way home in the evening, whilst waiting for my flight, I noticed a very important-looking man in a black jacket and striped trousers walking to the plane across the tarmac, followed by two minions carrying heavy briefcases.

Eventually we ordinary mortals were allowed on the plane and I sat next to a pleasant enough, rather nondescript, middle-aged man and chatted to him about this and that. After about half an hour we ran out of conversation, as one sometimes does, and I picked up the evening paper. The headline ran:

Richard Marsh, Joint Parliamentary Secretary, Ministry of Transport, visits Teesside.

'Look at this,' I said pointing to the headline. 'You know that chap who walked across the tarmac ahead of us with his retinue and is sitting a few rows in front of us. That pompous bastard must be Richard Marsh.'

'No,' he said. 'I'm Richard Marsh'.

Chapter thirty-five

'I Want One Now!'

Miss Daubeny (we called secretaries 'Miss' in those days) was my first secretary; I shared her with John Lewin, my office-mate at Gerald Eve & Co. John was much more senior to me so he had priority. Miss Daubeny was very young, very spotty, very shy, very nervous and very sensitive. So if there were several mistakes in a letter you didn't throw it back at her and say:

'This is bloody awful! Do it again, please' You said:

'Oh dear, Miss Daubeny, there seems to be a slight slip here, a terribly slight slip there, a tiny little slip here and a teeny-weeny one just there. Otherwise it's absolutely splendid, but I would be awfully grateful if you could possibly fine-tune it so its absolutely perfect.'

Miss Daubeny would blush scarlet, mumble 'sorry' and sneak out of the room with her head downcast. We both thought she was an absolute sweetie.

When I became a partner, I had a secretary all to myself, but she also took letters from assistants if she had a spare moment. I had an arrangement with Miss Richards, who took on the secretaries, that she would only give me pretty ones – and I must say she did a splendid job and never let me down. The reason I gave was that 'it reduced tension in me' – something for which the rest of my department was profoundly grateful. Before you get the wrong idea – or more likely after you've got it – let me stress that I always kept my secretary at arm's length, or, to be strictly accurate, slightly beyond arm's length.

The secret of keeping a secretary, indeed I would go so far as to suggest the secret of keeping any woman, is to make her feel needed. In this respect I started with one great advantage which I expect most of my readers do not possess: my memory was so appalling that I needed my secretary to remind me throughout the day what I was meant to be doing and with whom, and where I was meant to be going with what, with whom and to whom.

My last secretary, Allison Kirkness, whilst going up the M1 one Friday night, suddenly let out a screech. Her husband drew into the hard shoulder.

'What on earth's the matter?' he said.

'I forgot to remind Hilary to bring his dinner jacket with him to the office on Monday morning.'

That's what I call getting involved. How could she leave me when she knew I couldn't manage without her? She didn't.

On top of that, I used to ask my secretary to deal with my wife, my stockbroker, my insurance broker, my garage and my wine merchant. No secretary ever left her job through overwork but plenty do so because they are bored.

One measure which I adopted to make my secretary fully employed at all times was my 'serving hatch'. I used to have a serving hatch with a sliding window made in a division wall. I would then put my desk right against it and my secretary's desk slap up against it the other side. The result was that she never came into my room. When the mail came she would pass it to me and then I would pass it back to her saying yes, no, file, acknowledge, or I would dictate a letter. If either telephone went when I was dictating, one of us would close the sliding door and carry on working.

The increase in productivity was truly amazing but none of the partners followed my lead. They would call a secretary to their office each time they wanted to dictate and she would sit there for ages bored out of her mind, looking at her fingernails, while her dictator was merely talking on the telephone or to someone who had come into the room. However, some of the younger partners started to use tape recorders.

One secretary I had, Liz, who shall be surnameless, a seventeen-year-old in her first job, was so absent-minded that I had to remind her to remind me. That didn't work too well. Another, Caroline, went to Spain and fell madly in love with a Spanish guitarist. She kept on sending me telegrams asking for more and more unpaid leave to which I readily agreed – she was a very good, attractive secretary, but rather impulsive.

Eventually some old battle-axe in accounts said enough was enough and Caroline must either come back now or be dismissed. Caroline stayed there the whole summer. Her mother rang me up and said she was very worried. I said there was nothing to worry about as Caroline was having a marvellous time and was having much more fun than being my secretary.

At the end of the summer, Caroline brought her guitarist home to South Kensington where she lived with her parents. By the winter, the guitarist didn't seem quite so romantic and picturesque as they braved the cold on the icy pavements of South Ken, so she very sensibly managed to pack him off to Hollywood to seek fame and fortune there.

Some of the 'temps' I had were very colourful. One made straw hats packed with fake fruit on top and wore them throughout the day in the office. They must have weighed a ton – but presumably she wore them to advertise them to the other girls. It was summer and she looked very fetching, but I felt her millinery would have been more appropriate to Ladies Day at Ascot rather than in a rather scruffy office.

Then there was Julia, a very pretty, aspiring actress, whom I felt it my duty to take out to lunch so that I could tell my wife which were the best shows in town. I happened to tell Julia about a friend of mine who had arranged for the massive launching of a new product in Paris. When he arrived in Paris at the venue, his French organiser greeted him and said:

'Don't worry, Peter. Nussing is arranged!' He wasn't too good on his English negatives and positives

Julia thought this very funny and when she went to her evening job as assistant stage manager at His Majesty's Theatre in the West End, she tried it out. Soon, the whole cast, led by Dame Sybil Thorndyke, went rushing around the place shortly before the performance was due to begin, saying: 'Don't worry. Nussing is arranged!'

On one occasion my secretary was ill and I asked for a temp. Miss Richards said she was unable to find one, so I decided to take matters into my own hands and advertised in the evening paper, starting with the words: 'I want one now!'

Miss Richards told me that the headline in the advertisement had made people misjudge my needs; as a result the replies she received were from ladies who supplied personal service all right, but of a nature entirely unconnected with that of a secretary. In the circumstances, and bearing in mind I was a married man, she decided that to pass on the names and telephone numbers to me would be at once irresponsible and unwise.

Chapter thirty-six

Wining and Dining

'A man is in general better pleased when he has a good dinner upon his table'

Samuel Johnson

One morning at the office during Wimbledon Fortnight, a call came through for me from someone whom I will call Colin. He said:

'Are you doing anything this morning?'

Well, what would you say? I said:

'Absolutely nothing. Why do you ask?'

'Well,' he said, 'a client's dropped out and I've got two tickets for the Centre Court going begging.'

'Great,' I said, ' May I beg for them?'

I rang my wife who, like me, was doing absolutely nothing. It was arranged that she would pick up the tickets and meet me at Southfields Station. This she did and we parked the car with the car park ticket provided and wended our way to the appropriate hospitality tent. I said to my wife that we were on parade and had better be very polite and charming to all Colin's clients. As we entered the tent I looked around and to my great surprise I found I knew everybody – they were all from the St George's Hill Tennis Club of which my wife and I were members.

The 'hospitality' was all one could wish for; under the chandeliers a long table groaned with superb food, such as large prawns and lobsters, and the white burgundy was about the most expensive you can get – an aged Puligny Montrachet. (If you want to pretend to be a real wine buff, don't pronounce either t in Montrachet.)

There were several television sets scattered around the tent so that you could keep up with the play whilst wining and dining and, if you could tear yourself away, a ticket for the Centre Court awaited you.

We had an absolutely marvellous time and we expressed our heartfelt thanks to our host most profusely, but on the way home I began to ponder on just how all this luxurious entertaining had benefited the shareholders. Of course it didn't. At best it was just a perk for the Chairman, but I started wondering, and still do today, just how much of the vast amount spent on corporate hospitality is of any benefit to the shareholders. Is my experience typical? I rather suspect it is.

In Gerald Eve & Co., being a partnership, any expenditure on hospitality came out of the pockets of the partners, so we made quite sure it benefited us, the shareholders.

It has always been a tradition of the firm, started by Gerald, that we did not get work by entertaining clients; we got it by recommendation. When Royal Ascot was on, the big London firms of surveyors and estate agents were all at the races entertaining clients in their boxes. To my knowledge, no partner in our firm ever went to Royal Ascot during the week. If they did go at the weekend it was certainly at their expense and was not charged to the firm. Our hospitality was confined to asking a client to the firm's seminar, to a lunch or to a professional dinner. The all-time highlight of our firm's entertaining was the dinner we gave in 1980 to the legal profession to mark the fiftieth anniversary of the founding of Gerald Eve & Co.

I remember at one partners meeting chaired by TSD someone noticed an unusually large item of expenditure. Our accountant was asked to bring in the bill. He came in and read it out. It was for a number of lunches at the Mirabelle with TSD's client and friend who was a director of the Rootes

Group (Humber and Hillman cars). We all looked at TSD who grinned and said:

'You can't make bricks without straw.'

The conference room at 5, Queen Street was at the rear of the ground floor. I remember at one partners meeting one of the partners suddenly broke off in the middle of his sentence and said:

'Good Lord! Look at that large mushroom growing out of the corner of the room'

We all looked in amazement. There was a stunned silence. Eventually someone said:

'Anyone know of a good surveyor?'

The best professional function of the year was undoubtedly the Guest Dinner of the Rating Surveyors Association which was usually held at Claridges. On one occasion Geoff Powell and I took a few clients back to the Conference Room after the dinner for a drink. The senior partner, Eric Strathon, had, unbeknown to us, thought it advisable to keep the drink cupboard locked on that occasion, knowing that young partners might be on the prowl. It was therefore with some consternation and annoyance that we found we were unable to get to our drinks. Geoff disappeared and in seconds came back with what looked like a jemmy. Rather shrewdly, he handed the jemmy to me – who put it to good use.

We could find nothing but cognac in the cupboard, but that was not a problem. We had a great evening but I returned to the office next morning with an awful hangover. I was met by Fred Bishop, our chauffeur/butler who said, with a twinkle in his eye:

'Mr Strathon said there has been a burglary, The drinks cupboard has been forced open and all the brandy's gone. He wants me to go round the building and find out what else has been stolen.'

'Well, Bish,' I said. 'Kindly tell him it was me. He had locked the drinks cabinet and we were unable to give our clients drinks.'

'But Mr 'Ilary,' said Bishop, 'Mr Strathon said, "It *couldn't* be one of the partners."'

Oh, yes it could.

I think dear Eric found me a bit of a problem at times. I remember once he walked in to my office, found I wasn't at my desk and started to walk out again. On his way out he saw me sitting in the very corner – without a chair but as if I were on one.

'I expect you're wondering what I'm doing here,' I said.

'No, not at all, Hilpy,' he said gamely. He had long since ceased to be amazed at my antics.

'I am occupying one square foot,' I said, 'and am wondering what the remaining 149 square feet of my office are for.'

''Jolly good, Hilpy,' said Eric, 'Good luck. I do hope you find out.' He hurried out and, presumably shaking his head in bewilderment, returned to his office and sanity.

Our Conference Room, I always said, was the most important room in the office, for it was there that the partners lunched – those that were in the office, that is. I did not go to a university but I always said that the Conference Room was my university; it was there that we all took the tricky problems that presented themselves each morning. A brighter lot of people I have yet to come across and any sloppy thinking was soon pounced upon. I remember asking one question over lunch and one partner said:

'Hilary, it is easy to ask questions, but not so easy to answer them.'

Some indeed may not have been easy to answer but nonetheless there would always be some intelligent or witty comments. On one occasion at a partners meeting, one partner made a suggestion and I said:

'Over my dead body!' Back came one partner with:
'I think that could be arranged.'

I was the eighth partner when I joined the partnership but by the time I retired some twenty-seven years later, there were twenty-four. That meant we had to delegate decision-making so we formed a number of small committees, mostly with three partners on them.

You might think that the most important committee was the Finance and Policy Committee chaired by the Senior Partner, but in my view you would be quite wrong. Much more important was the Catering Committee, which provided luncheons for the partners – and the Car Committee which told partners what sort of car they could have. I took good care to make sure I was chairman of both. After all what could be more important than a good 'hot dinner' and a nice motor?

The chairman and members of each committee were elected annually, so how did I manage to get re-elected to my job year after year? Well to start with, as a mere glimpse at me would tell you, I have always been very fond of my food, so that was a good electoral advantage. Once elected, I made sure the partners got what they wanted by sending out each year a questionnaire to every partner asking what sort of food he would like for lunch.

As to the quality of the cooking, my motto was: 'complaints to me, compliments to the cook.' If anyone made a favourable comment on the food, I made sure he went to the kitchen after lunch and told the cook. This not only encouraged her to stay in her job – cooks can be a bit temperamental – but spurred her on to greater efforts.

As to the Car Committee, I ensured that it was the partners meeting who decided each year how much the various grades of partners could spend on their cars and how often they could change them. The committee might make recommendations but no more than that and, once the partners had decided, I made quite sure the committee had no discretion in the matter at all. So if a partner came to me and asked if it was all right if he bought car X which was £50 over the limit I would say:

'Absolutely fine, my dear fellow, but since I have no authority to exceed the price by even £1 it will have to go before a full partners meeting where, of course, I will give you my unqualified support.'

The partner invariably declined and chose a cheaper model.

We used to ask our clients to lunch with us; I felt this was a much greater compliment than taking them out to lunch or on some 'corporate hospitality' jaunt. There is no greater compliment than asking a client to a meal in your own 'home' and to meet all the 'family.' The clients also realised, (I hope) what a bright and knowledgeable bunch of chaps were advising them.

I usually used to give clients Berry Bros & Rudd's Good Ordinary Claret to drink with their meal – which I felt showed both discerning taste and sensible economy. A really expensive wine would make clients think our fees were unnecessarily high.

Throughout my time as a partner, wine was always served at lunch, but towards the end of my career I happened to be away on holiday during a partners meeting. When I returned I discovered that the partners, in a sudden fit of puritanism, had decided that wine would not be served at lunch unless clients were present. In fact, I think it made little difference because some partners merely poured themselves another gin and tonic before sitting down at the table.

Once we decided to ask the hierarchy of the Inland Revenue Valuation Office to lunch. That comprised the Chief Valuer and five Deputy or Assistant Chief Valuers. The six of them arrived at the stated one o'clock and got into the lift to go up to our Conference Room on the sixth floor.

Unfortunately, they omitted to look at the notice in the lift which said: 'LOAD NOT TO EXCEED FIVE PERSONS'. They were, to put it politely, a fairly stocky lot, and the lift was prepared partially to ignore their exceeding its weight limit and meet them halfway. And that is what it did. After that, halfway that is, it resolutely refused to move.

The Chief Valuer, Mr Christopher, announced fairly loudly to whom it might concern that they were stuck in the lift between floors. One of the partners went to find Harold Martin, the Office Manager, who was the only one who had the key to the lift room and knew how to wind up the lift. He had gone to lunch, so several people were sent out to see if they could find at which pub, café or restaurant he was lunching – or had he gone shopping in nearby Oxford St or Regent St?

It was half an hour before Harold came back, and fifteen minutes later the Chief Valuer and his cronies, thirsty and hungry – and literally wound-up – stepped out onto the sixth floor. By this time most of us partners were on our third gin and tonic so we gave them a pretty jolly welcome. There was laughter all round and after their harrowing and frustrating experience the gentlemen from the Inland Revenue soon caught up with us on G and Ts.

It was a very happy and rowdy luncheon party and I recall saying to the Chief Valuer that the art of negotiating with the Valuation Officer was to make him feel he was being unreasonable, because the typical civil servant likes to feel he is a reasonable man. As I have mentioned earlier, he said:

'I don't want my valuers to be reasonable. I want them to be right!'

Chapter thirty-seven

A Tale of Two Titles

One great advantage of having a Jewish client is that if you do a good job for him the word gets around – to other Jews. Whether they meet socially, in business, at charity functions, on charity committees or in the Synagogue, I know not, but the word certainly does get around. And the Jews have great respect for a professional man. Another businessman may try and take advantage of you – that's all part of the game, but a professional man puts your interests first. I must say I always found them delightful clients and most appreciative of our efforts. Most of my clients were Jewish and perhaps Mr Clore or Marks and Spencer, two very satisfied customers, had something to do with it.

One such client who was recommended to me by M & S was Lord Rothschild (3rd Baron 1910–90), one of the banking family. One of the reasons that the Jews got into banking was that under Protestant canonical law you were not allowed to be a usurer – which then meant simply 'a lender of money to strangers and charging interest'. So Jews were asked to fill this gap – in the confident knowledge of the Protestants that they would literally go to hell for doing so.

It is principally by moneylending that the Rothschilds made their millions – in particular by lending and delivering money to the warring parties in the Napoleonic Wars. Nathan Rothschild went one better than this. Flocks of carrier pigeons kept him apprised of the progress of the battle of Waterloo, which many expected Napoleon to win. So when he got news of the result he was able to buy up Government bonds at rock-bottom prices before the news of victory arrived by normal means.

Lord (Victor) Rothschild, father of Jacob Rothschild, was, when I met him, director of research at the Shell Oil Company. He was also from 1971 to 1974 director-general of the central policy review staff, Mrs Thatcher's 'think-tank'. Clearly he was a pretty smart cookie, as was his sister, Dame Miriam, who became the world's greatest expert on fleas. (Her Catalogue of the Rothschild Collection of Fleas at the British Museum was published in six volumes from 1953 to 1983, so there must have been quite a few different kinds of fleas.)

When I was shown in to Victor's office at Shell I was struck by how large it was – and what a huge desk he had. I felt some-what intimidated. However, that was not to last long because he immediately got out of his chair, came and shook me warmly by the hand, and waved me to a small table with chairs round it. We took our seats and were then on equal terms.

I was so impressed with this that I decided right then never again to meet a client whilst sitting behind my desk. As soon as I got back to the office, I ordered a table and four chairs, and for the rest of my career I talked to everyone in my office sitting at the table – whether it be a client or the office boy.

Lord Rothschild told me he had an orchard at his home near Cambridge and that normally he sold all his apples to M & S. However, there had recently been a hailstorm which left tiny lit-tle marks on the apples, causing M & S to refuse to accept them. So he was fed up with growing apples and would like to sell the orchard as building land. Could I help? M & S told him I was the best person in the country to do so.

This flattering approach was Rothschild's way of getting the most out of people, and I suppose it worked; certainly, I did my best to live up to my billing which I am quite sure he made up. However, I did find it rather unnerving.

I went up to see him at his house in Cambridge which at the time was the most expensive private house to have been built in

England since the War. He showed me round the orchard and then asked me if I would like a wash before lunch. I said I did and he ushered me to towards the loo saying:

'Don't be surprised if only water, not gold, comes out of the taps.'

After lunch we got in his car and he drove me to a small village he had had built nearby. The most important thing about it, he said, was that there was a wall on both sides of all the roads continuing right through the village. His architect had told him that this was very important because it made the village a single entity, gave it a oneness, and this instilled into the population a sense of community. I thought he had a good point. Today I recall all the walled towns I have seen in Tuscany and I think this must have had the same effect on the population there.

In due course I got him planning permission for the residential development of his orchard and then sold it for him to a builder. For some reason, probably because building land values were rising very rapidly at the time, my instinct told me that the purchaser would sell the contract on to someone else. I told my client this – and so it proved. His Lordship was quite happy about it, though perhaps he might not have been if I had not predicted it.

★★★

Just to show you that I hobnobbed with the aristocracy during my surveying career, the expression 'his lordship' which I have just used above has reminded me of an episode long forgotten. I turn from a tale of a mere baron to one concerning no less than a marquess.

If you are a mere commoner reading this, you may need me to explain to you that a marquess is three ranks above that of a baron, the intervening ranks being a viscount and then, one higher, an earl.

There are only thirty dukes, but thirty-five marquesses, the Premier Marquess of England being Lord Winchester who is the eighteenth of that name, the title being created in 1551. To give you an idea of their scarcity value (I hope I am not boring you, but I've only met one marquess in my life and I am going to make the most of this), there is in Whitaker's Almanac one page of marquesses, four and a half pages of earls, two and a half of viscounts and nine of hereditary barons.

Before you make a careful note of these statistics in order to quote them elsewhere with authority, I think I should warn you that they are derived from the 1995 edition of Whitaker's Almanac (127th edition, *J Whitaker & Sons Ltd* 1994) which I bought this year at our church fete for 50p. I think it pretty good value for money but, being nearly ten years out of date, I do wonder if (page 1052) the Uzbekistan Cabinet still has Mr Rakhmatull Makhamadaliyev as *Minister of Bread*.

The Marquess whom I am going to tell you about is that of Bristol who, before he died in 1985, was the sixth to bear the title created in 1826 by George IV. The Marquess said he wanted some advice on his estate in Lincolnshire and it was arranged that I would go and see him at his London home in Arlington Street, near the Ritz Hotel. I jumped into a cab and told the driver the address. I always talk to cabbies – the London 'black cab' drivers are the salt of the earth – but on this particular occasion I suppose it could be said that I was trying to impress him by name-dropping. I said:

'I'm going to see the Marquess of Bristol.'

'Oh yes,' he said 'Well don't ask him to show you his back.'

'Quite honestly, I hadn't planned on doing that,' I said, 'but what makes you say that?' I asked.

'Well, he was one of the Mayfair Boys, wasn't he? It happened before the War. It was quite a scandal and was in all the papers. He and a couple of other toffs lured a jeweller to a hotel in the West End, beat him up and stole all the jewellery he had brought to show them. They were all caught and sent to prison, but the reason I mention Bristol's back is that they were all given the "cat". You know, the cat-o'-nine-tails.'

I don't think the cabbie was too impressed at the people with whom I associated.

Ickworth House, Bury St Edmunds

On the day appointed to inspect his estate, I went to pick him up from his stately home, Ickworth House, near Bury St Edmunds, Suffolk. When we arrived at his estate in Lincolnshire, I decided it was in such a remote part that there was little demand for anything there, bearing in mind the prevailing economic climate. Unfortunately I could think of no bright ideas for making a fortune for him, apart from moving the town to somewhere that was somewhere.

My lasting memories are not of the remote, unattractive estate in Lincolnshire but of Ickworth House which I found delightful. According to my out-of-date guidebook, the house is open to the public in the summer and the park is open all the year round, but you would do well to check this if you plan to visit it. The mansion dates back to 1795 and has an immense, oval, domed rotunda. The guidebooks variously describe it as 'eccentric', 'amazing', 'unusual', 'Neo-Classical'.

One guidebook says the park was designed by Capability Brown, which is interesting because he died in 1783, twelve years before the house was built. By the way, Lancelot B. Brown, to give him his correct name, was known as Capability not because he was capable, but because he had the habit of saying that the grounds had capabilities. Well, he would, wouldn't he?

I left Ickworth House with an intriguing thought. Why on earth was there a photograph in the rotunda with the caption: *Ickworth House floodlit on the occasion of the visit of the Icelandic Minister of Defence.* What was the marquess doing? Gun-running during the 'Cod War'?

Chapter thirty-eight

Swept Paths and Fast Food

One day I found myself travelling up to Barnsley with the barrister, Keith Goodfellow. M & S wanted to stop up a highway so that they could make a service road for the delivery vehicles for their store. I was the expert witness whose evidence was meant to persuade the Planning Inspector from the Ministry that it was a jolly good thing to do.

You might think that delivery vehicles just wander up to a store and unload their goods. Not so. At this store there was a schedule of exactly when each lorry should turn up – and this included the small hours of the morning.

It's no good having a service road that's a cul-de-sac if you can't turn round at the end of it, so I had to explain to the inspector how much turning space was needed for the various delivery vehicles. You would be surprised how difficult it is to calculate the 'swept path' of various vehicles. I don't want to leave you trailing behind here so I had probably better explain what a 'swept path' is:

When a vehicle is moving forward and you turn its front wheels sharply the rear wheels do not follow the front – they take a short cut. When you are turning in reverse, the trailing front wheels swing out beyond the track of the leading rear wheels. Any of the ground over which the vehicle passes is called the 'swept path'. To make things more difficult, the swept path of an articulated lorry ('artic') is entirely different from that of an unarticulated one – as anyone who has towed a boat or a caravan will know. There is a formula for calculating all these swept paths and, if I could remember it, it would cover quite a few lines of this page.

Off the top of my head, it would depend on such things as the overall length and width of the vehicle; the distance between each pair of wheels; the distance between the back and front wheels and how sharply the wheels could be turned.

That's just for rigid vehicles. There would be different measurements to take and calculations to make for the artics.

I had to explain all this to the Inspector in the simplest possible terms, and to tell you the truth I rather enjoyed doing it. It is quite fun swotting up abstruse things and becoming an instant expert on them. A week later I had probably forgotten the lot. Today, in retirement, when I hold forth on a subject, I can be really impressive and convincing – as long as my audience knows precious little about the subject.

The Inspector sat late so that the inquiry could be finished in a day, and so it was pretty late when we caught the train back to London. Mentally exhausted after an extremely taxing day, I opened my briefcase, took out the latest James Bond book and settled down in my seat to read and probably dose off.

Counsel opened his briefcase and took out – a brief. He opened a fat bunch of papers to study, and indeed master, tomorrow's case. Some firms of solicitors just copy every bit of correspondence they have, instead of sorting the wheat from the chaff, and bung it at the barrister who then often has to spend hours wading through irrelevant documents to find those that matter.

The barrister then has to decide what line to take and what he is going to include in his opening address. I felt very sorry for him and, what ever my capabilities, glad I wasn't a practising barrister. Their clerks work them very hard and a barrister is lucky if he gets Friday and Saturday nights off. Sunday is often needed to prepare for a case on Monday, and Monday to Thursday evenings usually needed to prepare for the next day.

★★★

A few days later I went up north again – this time to The Wirral in Cheshire. I was acting for an American firm called G.E.M., the initials originally standing for Government Equipment Mart. It started by selling what we used to call 'Army Surplus' but gradually evolved into a chain of single-storey, out-of-town department stores. This was in the 1960s when out-of–town superstores and supermarkets had not yet caught on in Britain. The G.E.M. director sent over from the US to get these stores started was a Mr Wurtheim who was a delightful, cultured American who immediately took to the English way of life.

Unfortunately, through no fault of his own, he failed to get planning permission for any of these stores to be built and in all fairness, it was probably impossible to do so at that time. So, in true American tradition, just before the public local inquiry concerning his planning appeal in The Wirral was due to be heard, he was fired. He was replaced by a young, brash, crew-cut American who had never been out of his country before, and who seemed to have an unfailing facility for upsetting everyone he came across. His name was Mr Randolph Brevett – 'call me Randy'.

We all travelled up by train the night before and stayed at the Adelphi Hotel in Liverpool which probably sports four or five stars. There was the barrister, the solicitor, a landscape architect, a building architect, and Margaret Thomas and myself from Gerald Eve & Co.

Margaret was a geographer and her job was to explain to the inspector from which areas the customers would come, so that one could assess what effect the store would have on trade in neighbouring shopping centres. I remember Riley's Law of Retail Gravitation was relevant here. As far as I can recall, Mr Riley held that the larger the shopping centre the more people it attracts and the further they will travel to get to it. I will certainly go along with that, but I have a feeling Mr Riley rather cribbed the idea from our old friend Sir Isaac Newton.

As Margaret and I sat in the lounge of the Adelphi having a nightcap, a pageboy came round saying there was a phone call for Mr Eve. I got up and went to the phone booth, but someone got there just before me. I said:

'Excuse me, but I think that call is for Mr Eve.'

'Yes,' he said. 'Thanks.' He turned to the phone.

'Hullo.'

I tapped him on the shoulder.

'Excuse *me*,' I said, 'but that call is for *me*. *I'm* Mr Eve.'

'Gosh, that's a coincidence! 'So am I,' he said, 'I'll check which Mr Eve they want.'

He talked in to the phone:

'Which Mr Eve is the call for? There are two of us here.' Pause. 'OK, thanks.'

He turned to me:

'It's all right; it's for me, Mr H Eve.'

'You won't believe this,' I said, 'but I'm H Eve, too. I'm H M Eve.'

'That's ridiculous ' he said, ' because I'm H M Eve as well!' he turned to the phone again:

'I don't want to be awkward, but which H M Eve is the call for?' Pause.

'Thanks'. He turned to me:

'Unless you're Henry Michael Eve as well it's my call.'

'No, I'm Hilary Michael Eve. It's for you.'

I bought him a drink afterwards. As far as we could ascertain we were not related. He gave me his business card which I still have. He was General Manager of a firm selling sticky tape.

Next morning we drove off to the enquiry. In the car with me were the solicitor, the barrister and the client. I said to the barrister:

'Here's a tricky question for you. Do you know the difference between illegal and unlawful?'

Worried about being shown up in front of the solicitor and client as ignorant of one the finer points of the law, he muttered something about illegal being contrary to the law and unlawful being not in accordance with it.

'No', I said. 'Unlawful is against the law and illegal is a sick bird.'

The barrister laughed his head off, not at the joke, I fancy, but with relief at not being humiliated. After all, the joke is about as feeble as you can get – it was one on Tony Blackburn's I'd heard the day before on the car radio.

Shortly after we arrived, the inspector opened the inquiry and said he would take the names of those people who wished to appear. One man got up and said he was from the Communist Party of Great Britain and wished to make representations.

Our Randy was sitting next to me. He turned to me in amazement and said:

'Hilary, is that guy there a real live Commie?'

'Yes', I said, 'That is a real live Commie, all right.'

'And do you mean to say that that goddam' Commie is going to be allowed to get up on his feet and shoot his goddam' mouth off – and no one's gonna to stop him?'

'Exactly so. That's what these inquiries are for – to let everyone shoot his goddam' mouth off and get it all off his chest. It makes him feel very much better and tends to prevent violence. That's probably why we haven't had a civil war for over three hundred years.' I got a jibe in there.

In due course the inspector adjourned the meeting for lunch. In view of the dearth of eating-places nearby it was obvious we would be lunching at the Adelphi, so the inspector, a civilised man, said we would adjourn for an hour and a half.

All our party returned to the Adelphi and, except for Randy, took our places in the elegant, gracious dining room at a table next to the palms, thoughtfully booked by the solicitor, The solicitor passed the wine list to counsel (as is customary even though the solicitor – and ultimately the client – would be paying the bill) who savoured the list and suggested a white burgundy, a Meursault by a good shipper, would be a good wine with which to open the batting.

At that moment Randy turned up and instead of taking his seat said:

'Isn't there anywhere round here where one can get some fast food?'

His 'fast' rhymed with 'slayest'.

Counsel, as he should be, was the first to find his voice.

'Sorry, Mr Brevett, I didn't quite catch that. What sort of food was it about which you were enquiring?'

'*Fayest* food. You know. A big Mac.'

'A big what, Mr Brevett?'

'Mac, Mac! Jeez, don't anyone speak any goddam English round here?'

The solicitor thought he had better be solicitous:

'Mr Brevett, they are extremely accommodating here and I am sure if you explain to the head waiter precisely what you want he would do his level best to get the chef to accommodate your culinary requirements.'

Brevett looked round at us all to see if any of us had a glimmer of comprehension as to what he was saying. He saw none – we were rock solid.

'Oh, chucks', he said, 'forget it!' He sat down wearily, shaking his head.

'Someone order me a coke, for Chrissake!'

We lost the appeal, so I expect our Randy got the chopper, too – *fayest*.

★★★

Another tale of Cheshire concerns the late Ernest Marples MP. When Mr Keith Joseph was Minister of Housing and achieved the target of building 300,000 house a year, his right hand man was Ernest Marples who later became Minister of Transport and brought in Dr Beeching to make savage cuts in the railways in 1963. Marples was an old friend of my senior partner, TSD.

TSD had one of these intercoms that had a loudspeaker attachment so that if you pressed the appropriate key you could speak into the machine and the answer would be broadcast throughout the room. One morning Marples was in TSD's office to obtain his advice as to whether he should bid to develop a new shopping centre at Runcorn in Cheshire where a new town was being built.

'Runcorn, eh?' said TSD 'Just a moment, Ernie. I'll call our shops expert. He pressed a key on the intercom – my key.

'Hilary, what's the rental value of a standard unit shop in Runcorn?'

'Where?' I said, my voice booming around his room.

'Runcorn.'

'Never 'eard of it!' I said.

'Oh, well, would you come down, please.' He rang off.

Ernie said:

'Sounds like just the chap we want.'

Chapter thirty-nine

Traffic in Towns

For a good many years after the war, most of the cars produced in Britain had to go to export. To finance the war we had had to sell all our foreign investments and thus there was a severe balance of payments crisis.

However, after a while, cars became widely available for the home market and vehicular traffic increased steadily each year. One of the worst places for conflict between pedestrian and vehicle was the shopping street. In 1961 a book was published called *Traffic in Towns* which proposed that vehicular and pedestrian traffic in shopping centres should be separated,

The author of this book was Professor Colin Buchanan, a man ahead of his time, and his book had a great effect on current thinking about town centres. Buchanan was undoubtedly a very able man and revolutionised our way of thinking about shopping centres. However, he was so absorbed in his subject that he sometimes held forth upon it on unsuitable occasions.

One such occasion was a dinner dance of the Town Planning Institute which I attended. We had finished our dinner for some time before Professor Buchanan got up to speak. By this time we, particularly the ladies, were waiting to leave the table so that all the tables could be moved to the edge of the room and the dancing could begin. But before that could happen the Professor was going to have his say – and an unbelievably long say it proved to be. I found it incredibly boring and technical – and I was interested in the subject, so what the guests must have thought of it I can well imagine.

As the audience drifted into a torpor, I decided to liven things up a bit. I wrote on the back of my menu in big, bold letters APPLAUSE and on the other side HEAR, HEAR. I was sitting near a pillar and was thus able to hold up the menu without the speaker seeing it. After I had held up this menu several times quite a large section of the audience began to follow my instructions:

'We must separate the shopper from the traffic.'

'HEAR HEAR.'

'No longer must the shopper struggle in the wet and the cold on narrow, overcrowded pavements.'

'APPLAUSE.'

Looking back I suppose that if I had not indulged in this extremely ill-mannered exercise, the Professor might have cut his speech short, but frankly I doubt it. He was quite oblivious to his audience – (that is, until I livened it up) and assumed that they shared his interest in the technicalities of the separation of vehicular and pedestrian traffic.

For anyone involved in advising local authorities on the redevelopment of their central area, as I was, his book was required reading. In a further quest for knowledge, I attended a conference in the Midlands entitled *Redevelopment of Central Areas*. The principal speaker was an American and, when he had finished, someone asked him a question:

'Has anyone in the Unites States ever redeveloped a central area?'

The answer?

'Er...nope'

Loud and prolonged laughter.

It transpired that in America, when 'downtown' became obsolete, they just went and built another shopping centre out of town. Downtown' was left for the poor, which was usually the blacks, and the district often steadily deteriorated into a lawless slum.

I decided that if I were to advise councils on the redevelopment of their central areas I should get out of the office and go and see what had been done elsewhere. I went on visits organised by the RICS to brand new shopping centres outside Brussels and Paris which were very successful and exciting to shop in; but, unlike many others, I decided it was probably more important to go and visit the shopping centres that had failed. So off I went in the car with my assistant, Tony Taylor, to see what had failed and why.

Our first stop was at Gateshead, which is just the other side of the Tyne from Newcastle. It was a depressing sight. There was a multi-storey car park for a thousand cars which was nearly empty, and nearly all the shops were untenanted. The few traders who were there were pretty tatty. The whole thing absolutely terrified me. Supposing a scheme I got out ended up like that? It would be absolutely ghastly.

Why had the scheme failed? Because the multiple traders and department stores did not have confidence in it. Why was that? I decided it was because the centre was not big enough to stand on its own. I don't think I was alone in coming to that conclusion because, a long time later, another shopping centre was built at Gateshead, the Metro Centre – and this time it was the biggest in Europe.

Also, there was a lot of exposed, coarse concrete in the scheme which shoppers find most unattractive, boring and depressing. I think it's called 'neo-brutalism' and Churchill College, Cambridge is a good example of it. Personally, I think it's horrible but, whether you like it or not, it certainly has no place in a shopping centre. There is, of course, concrete and concrete. Some is made from really coarse, grey/black aggregate (which I think always looks ugly) and some is smooth and whitish and looks almost like Portland stone.

We went and looked at a few more new shopping developments in the North and Midlands and went home much wiser.

I must give a mention here of the Tricorn Centre at Portsmouth, built 1966 and now demolished, which was a horrific failure. Here again the car-borne shopper was presented on arrival with a neo-brutalist car park built of exposed, coarse concrete.

I understand the Royal Institute of British Architects gave the designer of the development, Owen Luder, a Gold Medal. This seems to me to be quite absurd because whatever one's subjective views about architecture, surely the first requirement of any building is that it is fit for the purpose for which it was designed; that it functions properly. And the Tricorn Centre did not function efficiently as a shopping centre. The reason it did not function as it should is that the development was not large enough to function on its own and was not properly grafted onto the existing shopping centre nearby in Commercial Road.

At the end of my travels, I made one or two resolutions about my giving advice on new shopping centres:

1. Bear in mind Riley's Law of Retail Gravitation – or as W Barrington Dalby said long ago about boxers: 'A good big 'un will always beat a good little 'un'.
2. It follows that, if you are going to build a new shopping centre, you had better either build a new town round it, like Crawley, Hatfield, Stevenage, Milton Keynes, or else build a very large centre with plenty of people living round it who live only a short drive away
3. If you are going to build shops next to an existing shopping street it is essential to graft it on. To do this, the new street or mall must be what I term 'a way of necessity'. For example, you might need to proceed along it to get from the car park or bus station to the High Street.
4. Ensure that the trade generators are in the right place. The Romans knew how to do this: in Pompeii you can still walk down what was once a shopping mall and find at either end of it there was placed a trade generator – namely, a brothel.

Today, John Lewis or Marks & Spencer might be more appropriate.

5. No neo-brutalism. The shopper is not to be exposed to concrete. If the frontage cannot be a display window, then shoppers prefer warm reddish brick.

6. Before committing yourself to building a scheme, go and see the important traders to ensure they have confidence in it. If they haven't, find out why. If things can be put right, do so, but otherwise don't build it.

7. Don't take risks with a shopping development. If you are in doubt, don't proceed. Confidence is a virtuous circle; lack of it a vicious circle. If the big boys won't trade there nor will anyone else – unless it's a small neighbourhood shopping centre.

The central area redevelopment which I devoted more of my time to than any other was Staines, which used to be in Middlesex but nowadays is in Surrey.

One day the Clerk of Staines Urban District Council (now Spelthorne Borough Council) came to see me with two of his colleagues. They spread before me a plan for the redevelopment of the town centre prepared by their consultants, Shingler Risdon Associates, and said they were very worried as to whether it would work. Would we advise then on its economic viability? I said we would be happy to do so, and in truth I was absolutely delighted to be given such a task which, immodestly, I felt well qualified to embark on.

I was truly amazed at the sheer size of the scheme and the extent of the land which they proposed to acquire by Compulsory Purchase Order. In particular, I noticed that Shingler Risdon proposed to acquire a number of shops in the High Street for what seemed to be no good reason, in which case a Compulsory Purchase Order would not be confirmed. Furthermore their calculations showed they were making a considerable profit by so doing.

To give an example, their figures showed they were going to acquire an old cinema in Clarence Street (a continuation of the High Street) for what it was worth as such, and then make a fat profit by erecting shops on it. But according to compensation law, if the cinema is worth more as a site for shops, that is the price you have to pay the owner. Also, as a generality, you do not make a profit by acquiring existing shops, because you have to pay open market value for them. You make a profit by acquiring back-land and opening it up and, in this case, creating new shopping frontage.

For these reasons, and more I won't bother you with, I decided that the scheme was hugely over-ambitious and its financial appraisal grossly over-optimistic. The best thing to do was to start again from scratch and get back to basics; I would use my experience of valuing shops, and also the lessons I had learnt and the resolutions made following my visits to new, failed, and successful shopping centres. My shopping centre would have everything going for it.

Staines High Street (see Plan) suffered from severe traffic congestion, particularly at its eastern end next to the railway bridge where there was a narrow junction with a service road on the south side. To alleviate this problem, the Council wanted to put a relief road behind the south side of the High Street and create a clockwise gyratory system, They also wanted a new bus station, some pedestrianised shopping and extra car parking.

There was a lot of back-land behind the south side of the High Street which extended to railway embankments. The question was: where would be the best place to site the relief road, bearing in mind the cost of acquiring the properties?

I decided on a method which I have not heard of being used elsewhere. I would put a value on every property between the High Street shops and the railway. To do this, I used an Associate in Gerald Eve & Co, George Gough. He was the ideal person for the job as he had had much experience in acquiring properties when he worked for the Greater London Council. In due course, George got out values for every property. I then got hold of an Ordnance Survey map at a suitable scale and marked in ink on each property its value as determined by George.

As soon as this was done it became quite obvious what was the most economical route for the relief road to take. I got out my felt-tipped pen (given to me by Peter Sellers, no less) and started to engage in what I termed 'Felt-tip Planning', with the benefit of a piece of transparent paper placed over the OS map. Some properties would be quite expensive to acquire, such as the Masonic Hall and the Telephone Exchange, for these properties could require 'equivalent reinstatement' under Rule 5 of the Acquisition of Land Act 1919, rather than be acquired at open market value. Thus, these two buildings still stand today, albeit overshadowed by multi-storey car parks and office blocks.

On my Felt-tip Plan, half way down the relief road on the north side would be the bus station. That would connect with the High Street by a pedestrian shopping mall which would emerge through a gap provided by the acquisition and demolition of two or three shops fronting the High Street. That mall would most certainly be a 'way of necessity'.

M & S had a store in the High Street which would be east of the junction with the new shopping mall. Would M & S like to back onto the new shopping development?

I went to see John White, M & S's Surveyor, and enquired if he was interested. Yes he was. I then asked him: if I altered the shopping scheme so that his store backed onto it, would he be prepared to pay a fair price for it? Otherwise I wouldn't bother. Yes he would. His word was good enough for me, so I altered the plan and made a junction in the shopping mall on the north side to extend eastwards as far as M & S. At the far end I put a small square and the entrance to a new multi-storey car park.

This car park would be quite expensive to build so, as we had quite a bit of land over, we would use the space as a surface car park, coupled with another new surface car park on land we

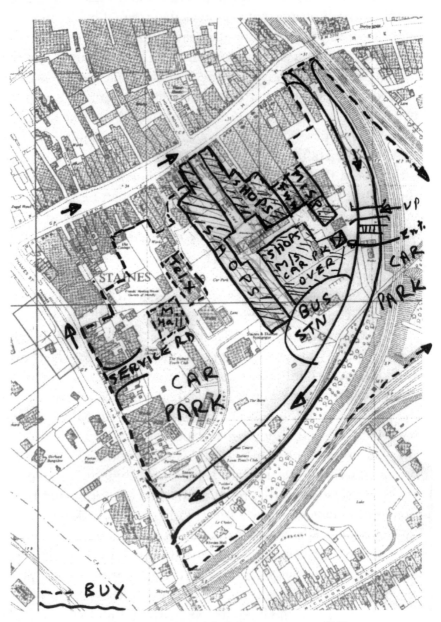

Author's felt-tip plan of Staines town centre, 1961

would acquire south of the mall. Then, later on, when there was a demand for it, we could sell the southern surface car park as a site for a block of offices and build a multi-storey car park on the northern car park.

One evening, Tony Taylor and I were due to meet the Council to explain the Felt Tip Plan to them and obtain their approval. We arrived early and I suggested we went to have a look at part of the land we were going to acquire which was the other side of the railway. We managed to climb over the fence of iron palings successfully and walked up the embankment to the railway line. Before crossing the railway I turned to Tony and said:

'Is this electric conductor rail still live?'

To my horror and before I could shout a warning, he leant forward and put his hand on the rail.

'No', he said.

Looking back on this incident I have a feeling that he knew something I didn't.

After viewing the land on the far side that was going to be made into a surface car park once the railway and its embankment had been removed, we returned to the iron palings. I climbed over them without difficulty but Tony was unable to get his leg up far enough as he was wearing very tight and trendy drainpipe trousers.

I climbed back up and stood on top of the paling and gave him a pull up. He got up successfully, jumped down and turned round to thank me. To his amazement I was standing quite still on top of the paling at an angle of about forty-five degrees giving a passable imitation of Eros atop the Shaftesbury Memorial at Piccadilly Circus.

This unique static tableau presented to Tony was made possible only by the fact that the bottom of my trouser leg had fallen on top of the long spike on top of the paling and was holding me up. The pose did not last very long. There came shortly a loud ripping noise as the stitching in my trouser leg got cotton fatigue and slowly came apart under the strain. I fell to the ground. When I stood up, shaken but not stirred, the side of my trouser leg made me look as if I was wearing a Susie Wong skirt – not quite the thing to wear when addressing the entire membership of the Staines Urban District Council.

We went to a shop nearby and, while I lurked outside in the shadows keeping the split side of my trousers close to the wall, Tony went in and came out with a packet of safety pins. He kindly attached the pins at regular intervals up my trouser leg and at least this temporary expedient stopped me from being arrested for indecent exposure.

The Council Meeting started with much jollity, both because of my appearance and due to Tony's amusing telling of the story behind it. With such a good start it was not surprising that my Felt Tip Plan was approved, as was the extent of the land to be acquired.

Compulsory Purchase Orders were made but before we made them there was one absolutely vital thing to do: obtain a letter from Mr Perry, the Urban District Council's Engineer and Surveyor. I met him and asked:

'Where is the worst congestion in the town centre?'

'At the junction of the High Street with the service road just south of the railway bridge.'

'Would you be prepared to give planning permission for any development served by this service road which would intensify its vehicular use?

'No,' he said.

'Do you think the Minister would support you on this on appeal?'

'Definitely,' he said.

'Then would you please be kind enough to write a letter to me to that effect?'

'Certainly.'

Armed with this letter, I felt confident I could acquire any property in the back-land at its existing use value.

We then started to acquire the various parcels of land from their owners and occupiers. I took the view it was not our job to try to acquire the land at less than its market value. The law said the Council should pay their ratepayers and owners the market value of the property so we didn't try to acquire them for less. As far as I can recall, we managed to buy all two hundred property interests at George Gough's valuation, but often we had to show Mr Perry's letter to the claimant in order to obtain a settlement at our valuation. In one case the claimant's surveyor asked for a lesser figure than George had placed on it. I consulted George and he agreed the compensation should be increased to his valuation.

The next job was to find a firm of architects who could bring to life the Felt-tip Plan I had prepared. We settled on Bernard Engle and Partners. I had seen some of their shopping developments, including the Brent Cross Shopping Centre and a new shopping centre over the Mersey at Stockport, and was most impressed with their work. Bernard Engle himself was quite charming, had long, bushy, white hair around his head and looked like some classical composer such as Beethoven. He was the aesthetics man and had surrounded himself with some very competent, commercial architects – a winning combination. I showed him Shingler Risdon's plans, sections and elevations. As he looked at them he remained silent. Then, almost imperceptibly, he started to shake his head very slightly. When he had seen them all he looked up at me, put his index finger and thumb together and pushed his hand out towards me, and said in a strong Central European accent:

'No music!' Pause. 'No music!'

I though that was absolutely marvellous.

Bernard Engle and Partners approved of my Felt-tip Plan and readily agreed to have warm-coloured brickwork rather than exposed concrete in the non-retail areas that shoppers would pass by. The warm red brickwork looks attractive today whereas concrete would look depressing – and pretty dirty by now after twenty-five years.

After a while, the plans were prepared and we were all set to go and seek a developer, except for one thing – the property market was going into deep recession and no developer wanted to build a shopping centre or an office block. How long this lasted I cannot recall – it certainly seemed a long time – but after what seemed an age we had an approach from the Duke of Westminster's Grosvenor Estate Commercial Developments Ltd ('GECD').

I was absolutely delighted because I had seen one or two of their new shopping centres which I thought were excellent. In particular, I admired their development at Chester (completed 1965) where their shops had integrated and blended splendidly with the famous Chester Rows.

However, the Council was required to offer the site on the open market and asked us to prepare a press release. Over thirty developers responded to it. We selected a short list of nine and got out a schedule of information about all the companies together with their 'track records'. (What the difference is between a 'track record' and a 'record' I have yet to discover, but the term has become fashionable). I was pleased when the Council decided to choose GECD – 'because it had the best track record'.

The work went ahead and when it was finished in 1979, Her Majesty the Queen was asked to come and open the new shopping centre, which she did, accompanied by the Duke of Edinburgh. It was about five o'clock by the time Her Majesty had finished her inspection, – she had had another appointment in the morning – and she had a mere fifty yards to walk, after a very tiring day, to get to her car at the end of the new shopping mall. Both sides of the mall were packed with schoolchildren and I swear she must have taken half an hour to get to the car, as she took so much time to chat to the children. I was most impressed with her professionalism and thought how very lucky we were to be blessed with a queen who was so devoted to her job and took an infinite amount of trouble to do it to the very best of her ability.

At the reception, to which the Council was kind enough to invite me, were, amongst others, the Duke and Duchess of Westminster, representing the development company, and also, I was pleased to see, David Wynne, the sculptor; he had made the most beautiful sculpture and fountain, *The Five Swimmers*, which was placed in the middle of the square within the shopping mall. (Amazingly, it is no longer there.)

I was at school with David – he shared a study with my brother – and I have no doubt that he is the best sculptor in the country, even if he is not, perhaps, the most famous. You can see a number of his sculptures in Tresco, one of the Scilly Isles, but his best known and best loved sculpture is, perhaps, *Boy with a Dolphin*, 1975, in Cheyne Walk by the Thames embankment near Albert Bridge, London.

The Staines scheme extended over twelve acres and comprised 250,000 sq ft (roughly 25,000 sq m) of shopping, with two large stores, 45 standard units, a 30,000 sq ft (3000 sq m) two-storey extension to M & S, and parking for 1100 cars.

This account of how it was brought to fruition has been grossly over-simplified; I have said nothing of the years of delay, frustration and expense occasioned by having to wait for Government decisions on the Comprehensive Development Area and Compulsory Purchase Orders, nor of a previous developer who fell out when the market became depressed. It was sixteen years (1961 to 1977) from the date we were instructed to the date GECD started the development, the centre opening on 11 October 1979.

Later we sold or leased as sites for office blocks, shops etc, at a great profit to the Council, the adjoining land we had bought and used as surface car parks.

Today, the High Street is pedestrianised and a relief road north of it has been built which skirts the new Two Rivers shopping centre which, like the Elmsleigh Centre, is connected to the High Street by a shopping mall.

It gives me great satisfaction now to see that the shopping centre at Staines is attractive and flourishing; and also to think of the part that my firm – George Gough, Tony Taylor, and myself, that is – played in bringing it about. Apart from the redevelopment of the central area, we helped to make our clients, the Staines UDC, several million pounds.

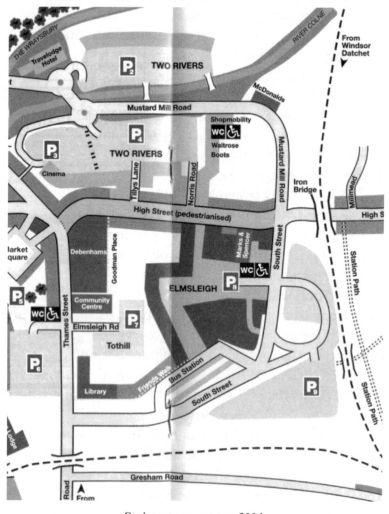

Staines town centre, 2004

Chapter forty

An Uncooperative Co-op

From a job acting on behalf of a Council, I move to one where I acted against one. The Royal Arsenal Co-operative Society owned the freehold of a retail furniture store in the centre of Woking. They told me that they had a Compulsory Purchase Order made on the store owing to their being included in a scheme for central area redevelopment. They had been offered a store on lease in the new development but were anxious to retain their freehold. Could I help?

I went and had a look at the site and then studied the plans for the new shopping centre. I thought it an excellent scheme so it seemed to me there was little I could do to substantiate an objection to it. For want of anything better to do, I thought I had better read all the bumf concerning it to see if it could give me any inspiration.

I found that the strategy underlying the whole scheme was the formation of what was called 'a motorway box'. This 'box' was enclosed by two motorways and two A roads forming four sides of an approximate square. Inside the square was to be the enlarged Woking shopping centre. I decided to challenge the whole concept of the 'motorway box'.

After consulting my clients, I wrote to the Council and the developers to say that not only did we object to the Compulsory Purchase Order on our client's land, but we also objected to the whole scheme on the grounds that the basic concept of a 'motorway box' was flawed.

Not only was ours the only objection to the Compulsory Purchase Order, but it was the only objection to the plan as a whole. It could mean the Minister might order a public local inquiry into both the scheme and the Order. All this would take some considerable time, and both the delay and the inquiry would cost a lot of money.

The Clerk of the Council called a meeting which was attended by the developers, my clients and me. After some discussion, to my great surprise the Clerk asked me to draw up heads of an agreement between the Council and my clients. I wasn't quite prepared for this 'surrender' so asked for an adjournment of five minutes.

I returned to the meeting and asked for terms somewhat on the following lines:

1. My clients would be given the site in the central area edged pink on the attached plan.
2. The site would be an unencumbered freehold.
3. The Council and the developers would use their best endeavours to help us obtain planning permission for a retail furniture store on the site.
4. Our clients would use their best endeavours to build the retail store on the site as quickly as possible.
5. As soon as the new store was built, my clients would move in to it and give up their existing freehold site to the Council, thus ensuring continuity of trading.
6. The Council would pay my clients for 'disturbance' caused by the move, in accordance with the statutory compensation provisions.
7. The Council would pay my clients reasonable legal costs and surveyors fees.

The Council Officers and developers withdrew and return within a couple of minutes; they accepted our terms in their entirety. I like to think my clients and I did not betray any sign of surprise or delight on our faces as we shook hands and departed. As Winston Churchill said: 'Magnanimity in victory'. But I must confess that as soon as we went round the corner of the Council Offices we all danced with joy and hugged each other.

Not only was I overjoyed at my clients getting exactly what they wanted – more than they ever dared hope for – but I was very relieved that I did not have to go into the witness box at a public local inquiry to put forward a plausible alternative to the 'motorway box'.

Looking back on the whole matter today, I feel this tale is the old story of throwing money down from the balcony to the street musician to get him to play somewhere else – so that you can get on with your activities in peace.

Another Council I was asked to advise was Farnham Urban District concerning their town centre in which was a cattle market that had recently ceased trading. The proposals I suggested involved demolishing two pairs of Victorian semi-detached houses. This caused a local uproar and, shortly after, I found myself dining at the Surveyors Club Guest Dinner seated next to John Betjeman. He was, with Nicholas Pevsner, one of the thirty founders of the Victorian Society in 1958. I mentioned to him the opposition that had arisen over the proposed demolition of these two Victorian houses. He was aware of the proposals at Farnham and I expected him to object strongly to the demolition, but he did not:

'Oh no,' he said 'I certainly don't object to their demolition. They're just Victorian tat.'

I was very surprised at what he said but impressed at his discrimination. I said:

'Well, why then do so many people object to their demolition?'

'Because they are worried about what is going to be put in its place. That's why.'

As you can see, I have never forgotten what he said. I have found it so often to be true, but I don't think he went far enough. People in general do not like change in their neighbourhood. They would like to stop the clock. But they didn't stop the clock in Farnham; the 'Victorian tat' was demolished and the site used as part of an extended shopping centre.

I will give you an example of resistance to change; in Bosham, near Chichester, West Sussex, where I live, a shipyard closed as it was no longer economic, and a planning application was put in for a development of about two dozen well-designed terraced houses and a boathouse in a little enclave. It raised a storm of protest among the older inhabitants. The protest was not on the grounds that the change of use from industrial to residential would result in loss of local jobs – which would have been a fair point. The protestors simply resented change and said it would be 'detrimental to the amenities'. As a result this attractive development never took place and there remains today on the site the ugly, former boatyard building with a large, rusty, corrugated iron roof, overlooking Chichester Harbour. A veritable eyesore and highly 'detrimental to the amenities'. Suppose there was nothing on the site and someone proposed to re-erect an 'ugly, former boatyard building with a large, rusty, corrugated iron roof', there would be an even bigger outcry, but in effect that is what was re-erected, rather than demolished.

There's a lot of mud around in Bosham and quite a few people are stuck in it.

Chapter forty-one

'A Very Clubable Man'

'Boswell is a very clubable man.'

Samuel Johnson 1709–84

One of the joys of being a chartered surveyor, I found, was being a member of a surveyors dining club .One of these clubs, the 1894 Club, was started, not surprisingly, in 1894 of which my father was a member, and later, my brother, Adrian. My first cousin, Douglas Trustram Eve (obituary in Appendix 2), put me up for the 1924 Club where I have enjoyed being a member since 1955. The '24 Club having been started thirty years after the '94, it was thought a good idea to start a 1954 Club. This occurred and was duly followed thirty years later by the creation of a 1984 Club.

We meet at a London Club in Pall Mall three times a year. On one occasion I was sitting next to a member, Holroyd Chambers and he told me that he and his wife both suffered badly from 'slipped discs' and their life was a misery. I told him to go to an osteopath but he said he had been warned against them and had no intention of doing so.

Using what I like to think are my considerable powers of advocacy, by the end of the evening I had persuaded him to go to Marianne, my Belgian lady osteopath in London. Both he and his wife went to her and the next time I saw 'Roy', as he liked to be called, at dinner he told me I was the most wonderful person on earth. I had persuaded him against all his lifelong convictions to see an osteopath – my Belgian lady. She had put them both right and revolutionised their entire lives. That was entirely due to me and he was eternally grateful.

Since I was such a splendid fellow, he was going to put me up for the Surveyors Club. I was duly accepted and found most of its members were architects. I think the reason for this is that quite a while ago the term 'architect' was virtually synonymous with 'surveyor'. For instance, in 1669 Charles II appointed Christopher Wren the Royal Surveyor. The King had previously asked him to survey the defences of Tangier (which he refused) and the old St Paul's Cathedral. Also, in my day, one distinguished architect member had the appointment of what I think is correctly described as Surveyor to the Fabric of Westminster Abbey.

The club was started in 1792 and contained both distinguished architects and descendants of distinguished architects. Roy himself was descended form Sir William Chambers who designed Somerset House in London. Other descendants included those of Sir Alfred Waterhouse (Manchester Town Hall; Prudential Assurance, Holborn; Natural History Museum, South Kensington) and Sir George Gilbert Scott (Albert Memorial; Foreign Office; St Pancras Station and Hotel). There were also a few chartered surveyors, who were a lot more distinguished than I was, but I qualified for membership as the descendant of a distinguished chartered surveyor.

The method for electing the officers of this club was based on the 'It's Buggins's turn' principle. The result was that within a year or two of joining the club I became Buggins. The first thing that happened to me is that the outgoing Buggins dumped several large, rusty, tin trunks and boxes on me at the office. These contained the club records and Minute Books since the club started in 1792. The members dined pretty well a hundred years ago, as revealed by a (slightly damaged) menu card I possess, reproduced here. This dinner was held at the Albion Tavern in Aldersgate Street in the City of London.

A cursory glance at the old Minute Books showed that they would make fascinating reading – if I had time; but not only had I to practise as a chartered surveyor but now I had to organise six dinners a year including a guest dinner; help the President organise the outing for the Annual Summer Meeting; send out all the invitations to members for each event; deal with all the replies; book the venue, choose the menu and the wines; make last minute alterations to numbers due to cancellations; collect all the subscriptions, write and distribute the minutes; ensure every member had submitted a photograph of himself, and issue a revised list of members printed on linen (Why linen, I hear you cry? Why, indeed?).

The job of Hon Sec must have been even worse when the club started, because the club used to meet monthly for the purpose of conducting business and settling disputes.

Thank you, Roy Chambers.

The Rating Surveyors Association (RSA) was a splendid Association to join. They had two dinners a year: the first was a Members Dinner and the second a Guest Dinner. The Guest Dinner was usually held at Claridges and there were invariably many lawyers there, one or more of whom could be relied upon to give a really good speech. I remember on one occasion Lord Justice Singleton was the speaker and he told a very funny joke that made everyone laugh. As the laughter

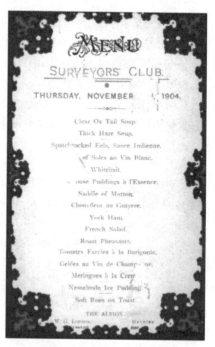

Menu for the Surveyors Club Dinner at the Albion Tavern, Aldersgate Street, City of London, 1904

died down, the old boy next to me leaned towards me and said in my ear:

'He told exactly the same joke here twenty years ago.'

The President also got asked to one or two other dinners, like the Rating and Valuation Association.

The RSA wasn't just about hot dinners; it was a great help to members. It also advised the Government about rating matters when it was asked to – and sometimes when it wasn't. I have a letter from my cousin, Douglas, saying that I should join the Association in view of my father's long connection with it. I duly applied to join, was elected, and received the List of Members.

I soon realised that my father was not the only Eve to have had long connections with the Association, which was started in 1909. On page 2 of the List were the names of the Past Presidents:

1919 Sir Herbert Trustram Eve (uncle)
1922 W Harold Eve (great uncle's son)
1923 C Gerald Eve (father)
1934 Frank N Eve (great uncle's son)
1944 J Douglas Trustram Eve (first cousin)
1949 H Brian Eve (great uncle's grandson).

All the above were descended from my great-grandfather, Richard Eve. Reading that list made me want to become President one day, so after a while, I managed to get elected to the Committee. Twenty-five years after I became a member, I was elected President, so I could add to the above list:

1980 Hilary M Eve

There were also a number of partners from Gerald Eve & Co who became Presidents:

1931 C D Conquest
1954 Tom S Dulake

1956 H W Jack Powell
1958 Eric C Strathon
1959 James P Lambert
1964 P G E (Bill) Haddock
1978 W Martin Hattersley
1984 Alan J Duncan
1986 Michael St J Hopper
1995 Peter R Jones TD

Being President of the RICS was very nearly a full-time job and involved travelling almost non-stop all over the country and abroad. Indeed, three partners of Gerald Eve & Co have had the job: Gerald, Eric Strathon and Jeremy Bayliss. But the Presidency of the RSA involved very little work. You had to chair a few committee meetings, had a seat at the top table (at the very end) of the RICS Guest Dinner, chaired the Members

26th Annual Luncheon of the RICS Rating Diploma Holder's Section in
1980. *(Left to right):* Alan Kennedy, Chairman; J N C (Jimmy) James,
President, RICS; and the author, President, Rating Surveyors Association

The author, when President of the Rating Surveyors Association, presenting the prize for the Special Diploma in Rating to Mr Abernethy

Dinner (no speeches). You also attended the lunch and presented the prize to the winner of the RICS Special Diploma in Rating. Your biggest day was presiding over the Guest Dinner, at Claridges in my year of office, at which you had to make a speech proposing the health of the guests.

Shortly before I was due to make my Presidential speech, I met my friend, John Guillaume, whilst travelling home on the train. There was no one else but us in the carriage and John asked how I was getting on with my speech. I said I had it on tape and had a small tape-recorder with me, so John asked to listen to it.

He had recently been Master of the Solicitors Company and had had to make several after-dinner speeches, so I was interested in his views. He listened to my speech for a short while and then switched it off and said:

'You don't pause in the right places.

'Come on,' I said, 'I've got to stop sometimes to think what I am going to say next.'

'Oh no you haven't, Hilary,' he said. 'For an occasion like this you don't stop to think; you have to have it all written down'

'But that's not my style, John. I prefer to speak from notes.'

'Never mind your style up till now, Hilary. This is a big audience. You will have a lectern in front of you and almost no one will know whether you are reading the speech or not. [That proved to be true.] When Churchill wrote a speech, he would put a vertical line on the typescript whenever he decided there should be a pause. You should do the same. I know you are used to conducting cases at Local Valuation Courts, but this is different, To make people laugh at your stories, your timing and intonation must be perfect.

'By the way, go and visit Claridges and find out where you will be sitting, ask them to put the lectern in place and see how much light you have on it. You might need reading glasses. Such a visit helps to familiarise you with the environment and calms the nerves.'

I took all his excellent advice. My guests seated either side of me at dinner were Nigel Bridge, on my left, who by then had become a Lord of Appeal in Ordinary (a 'law lord') and Sir Thomas Lund, Secretary-General of the Law Society, whom I had heard give an excellent, witty speech at a Solicitors Company dinner to which John had invited me.

At the dinner, I decided that since for the first time in my life my speech was all written down, I would not be abstemious but would drink the same amount as the members and guests. That way I would be relaxed and in tune with them when I spoke. So while my guests either side, each of whom had to speak, toyed nervously with their drinks, I imbibed and kept up with the rest of the diners.

My theory was that a nervous speaker would only attract nervous laughter but that a confident speaker who was clearly at ease and enjoying himself would spread his enjoyment to his audience.

On the night, my theory worked and John's advice was impeccable. I thoroughly enjoyed making my speech and I was so relaxed that some people jocularly barracked me, as sometimes happens at, for instance, wedding receptions, and I joined in with some off-the-cuff repartee –which was just my style. I could divert from the text of my speech, when I felt like it, for an aside, because I knew I had a text to return to. Thanks to John's tip about timing and intonation, my jokes seemed to go down well and I felt that the audience enjoyed it as much as I did.

I have always enjoyed public speaking, and I think if you can make people laugh once more in their lives then they would have done otherwise, that is an achievement. The first time I can recall speaking in public, apart from official debates at school, was at a Young Conservative meeting at Weybridge when there was a debate on the Old Age Pension, as I think it was called at that time. The debate was getting rather boring and desultory and the chap next to me said:

'Hilary, this debate is dying on it feet. Can you get up and say something to get it going.'

'OK,' I said, and after thinking for a moment or two I stood up, was called by the Chairman and said:

'I think old people are a drag on the economy and that everyone over sixty should be shot.' Then I sat down again.

Several young girls, who had never in their lives spoken in public, jumped indignantly to their feet. One was asked to speak:

'My father is sixty-two and all the family love him. I think Hilary's suggestion is the most ridiculous and unkind suggestion I've ever heard...'

I have to confess that that was not the last time that I have made a deliberately provocative remark in order to stir things up. However, my partners were pretty good at dealing with me in debate at partners meetings. My brother, Adrian, when Joint Senior Partner with Geoffrey Powell, used to chair the meetings and thought that one of his duties in life was to stop his younger brother from being too bumptious. Unfortunately for me, no one seemed to object. Frequently, when we were discussing some matter at a meeting, I would come up with an idea (I told you earlier that I thought of myself as an 'ideas man'). Adrian, in the Chair would typically say:

'Well, we've heard your views, Hilary – and pretty stupid they are, too. Has anyone got any sensible suggestions?'

There would be laughter all round, the partners feeling that I had a sufficiently buoyant personality not to need rescuing, and that perhaps Adrian was doing a job that was needed from time to time. Perhaps.

I remember saying to my partners once after a somewhat lively debate:

'One of our difficulties is that Adrian and I can never agree.'

Adrian said: 'I don't agree!' and, when everyone started roaring with laughter, he looked around in bewilderment.

Back to the RSA and its Annual General Meeting. I was asked in April 1986 to propose the Vote of Thanks to the outgoing President, Geoffrey Keeling (1929–2004), who hailed from Manchester. I decided to thank him in verse. Let me set the scene for you:

Maggie Thatcher is Prime Minister and has said in her election manifesto that she will abolish rating. Kenneth Baker is Secretary of State for the Environment and is thus the minister responsible for rating.

The RSA members are worried they might lose all their rating work, so Geoffrey undertakes personally to make representations to the Government:

1. Tell them rating is 'a good thing', not just for us, but (fortunately) for the country as a whole.

2. Tell them *why* rating is 'a good thing'.
3. Tell them therefore, that the national Rating Revaluation, which was to be in 1983 but had been cancelled, should now go ahead.

Geoffrey, bless him, succeeds in his mission and I get up to propose a Vote of Thanks to him for his year of office, which I deliver in a broad Mancunian accent (I hope):

> Our President, Geoffrey Keeling
> Had an extraordinary feeling
> That, though Maggie said rates
> Were one of her hates,
> We'd probably carry on dealing.
>
> But just to make doubly certain
> She didn't ring down Iron Curtain
> On what for us nutters
> Were our bread and butters,
> And some put their very last shirt on,
> He got pencil and masses of paper
> And wrote to one, Kenneth P. Baker,
> To say: 'Stop debating
> This question of rating,
> And if Maggie won't, jolly well make 'er.'
>
> It began: 'My dear chummy, Kenno,
> Please shufti this 'ere enclosed memo.
> If you cut out the rates
> We'll no longer be mates
> And we'll make a disorderly demo.'
>
> So, Kenny thought: that's not so great,
> Upsetting old Geoffrey, my mate.
> So to Ten Downing Street
> Our Maggie to meet,
> He went off in a terrible state.

He said: 'I've put cold towel round my head,
And considered it all night in bed.
I've pondered in Church
And, after ten years' research,
We can't think of a system instead.'

Maggie sighed – ever so fleeting -
And went into Cabinet Meeting.
She said: 'You lot, listen, now, then!
I've just been convinced by our Ken
That rates really do take some beating.
I knows in my manifesto
I said I'd scrap rates thus – Hey Presto!
But we've got to face facts
And rates as a tax
Are hereby declared much the best-o.

'So we will have rates revaluation.
It's all for the good of the nation.
Oh, and I nearly forgot,
It'll sure help a lot
The Rating Surveyors 'ssociation.

I turn to Geoffrey, seated next to me:

So, Geoffrey, in your term of office,
With the help of innumerable coffees,
For good not for evil,
You've got us a reval.,
So you've earned this – a large bag of toffees,

And with that, I hand him a family-sized bag of Quality Street toffees – a somewhat meagre reward for what he had achieved – but a heartfelt token of our appreciation.

Chapter forty-two

'A Nation of Shopkeepers'

'L'Angleterre est une nation de boutiquiers'
Napoleon at St Helena

There's no doubt that the subject on which I spent most time in my professional career was the rating of shops. I was fascinated by what attracted people into shops. I would walk along Oxford Street and find one ladies' fashion shop chock-a-block with people and another one, which looked similar to me, completely empty. Was it the display window, or lack of it, was it the music or was it what they knew they would find inside? It always amused me to see in a parade of shops someone, such as a butcher, who had no customers, standing in the doorway of his shop, with his arms folded, aggressively blocking the entrance to his shop, and wondering why nobody was coming in.

There is great consumer resistance to coming into a shop, which the shopkeeper has to remove by various methods. The most important one is for the customer to feel he will not be confronted by the shopkeeper saying as soon as he comes in, or implying by his presence:

'What the hell do you want, coming in like that without so much as a "by your leave" or "if I may?"' or

'If you enter my shop don't you dare go out again without buying something!' or,

'If you come into my shop I am going to pester you until I make you buy something you don't want.'

The most annoying habit I find is when I shop abroad and the shopkeeper follows you around the shop wherever you go. If you pause for a moment he will come out with what he thinks is a good selling point like:

'Mister, this wallet is genuine artificial leather.'

The department stores, and what were known as 'chain variety stores' (M & S, Woolworths, Littlewoods, BHS) did not have this problem of overcoming customer resistance, as their retail format was more like a market with stalls and rails of clothing.

However, the vagaries of the various shopkeepers and their selling methods, fascinating as they might be, were not applicable to shop rating, for rating surveyors had to value on the hypothesis that the shop was vacant and to let in the open market.

Having said that, my job was to find the correct rateable value of a shop which, almost invariably in my experience, had a rental value higher that the rateable value assigned to it by the VO in his Valuation List. The reason for this was the time lag between making the valuation and it coming into force. During this intervening period, rents rose – due sometimes to increased sales but mainly due to inflation. I will deal with this problem at some length later on.

The first thing I had to decide was how to spell 'rateable'. Having read the statutes and learnt the definition of *rateable value* off by heart, I was more than a little upset to find that *The Times* newspaper spelt rateable 'ratable'. So I wrote to *The Times:*

> 'Sir,
> Why does *The Times* spell rateable
> In a way that is highly debateable?
> I would think it compatible
> If the sound was thus: rattable,
> But to leave out the 'e' I find hateable.'

I got a reply. It said:

'<u>Private</u>, and Dear Mr Eve,

A trick that we keep up our sleeve
Is to leave out the 'e'
When it doesn't sound, see?
But we really don't aim to aggrieve.'

I find that Oxford University Press, which is part of Oxford University, comes down firmly in favour of 'rateable', but *The Times* has decided on 'ratable'. I have concluded therefore that he in the Government who decides on the spelling to be adopted in Acts of Parliament went to Oxford. As you will have noticed, I have adopted, and will continue to adopt, the spelling in the statute.

Don't think for a moment that only the spelling of rateable is debat(e)able. A siz(e)able mob has unshak(e)able views on how to spell sal(e)able. Personally, I think 'ratable' is liv(e)able with, but not lik(e)able. Shall we leave it at that?

Whilst we're on the subject of *The Times* newspaper, it is a good, and probably the only, opportunity I will have in this book to tell you about the letter I wrote to *The Sunday Times* to which I had a personal reply. On the back page of the *S.T.* was a heading: **'Forecast: Mediterranean'** and, underneath, it gave the expected temperatures and weather for various resorts. Included among the Mediterranean resorts were, amazingly, Agadir, Algarve, Canaries and Madeira; so I wrote to the paper as follows:

'Sir,

Having read the back page of your journal
I'm a really disgusted old Colonel.
For you highly intelligent primates
Under 'Mediterranean' climates
Include...The Canaries.
Also down there is
Madeira
(though nearer).

Agadir
Is quite near
But, let it be said
Still not in the Med.
The Algarve is the nearest,
But what is quite clear is
'Tis not just my romantic notion
That has them in the Atlantic Ocean.
So, to make a Colonel happy
Please…
Buy an atlas; make it snappy.

Yours faithfully, but not too seriously.'

Two days later, back came the reply:

'Dear Disgusted old Colonel,

Of course you are right,
But please don't get frantic,
You are still in the Atlantic.
We are forced by tradition
Under threat of sedition,
To follow Met Office law.
Though away with the fairies,
This leaves the Canaries
High and dry,
In the Med…What a bore.

Yours sincerely

Keith Austin
Acting Letter Editor'

Exactly two months later, I looked on the back page of my *Sunday Times* and, bless my soul, there was a new heading: **'Forecast: Resorts'**.

Not all of us can boast of altering the format of a national newspaper, so I think after all I will take my place in history.

I really must now get back to the matter of shop rating which this chapter is meant to be about.

As I mentioned in Chapter 10, we had national Rating Revaluations in 1956; also in 1963 and 1973. The 1973 list was published in December 1972 and came into force on 1st April 1973. You had a year to make an appeal if you wanted any reduction you might get to date back to 1st April 1973.

You made an appeal by asking the VO for a proposal form. In due course he would send you the form with the 'Description of the Hereditament (e.g. 'Shop and Premises') and the present assessment written on it. There was a space on the form for you to state what the grounds were for a reduced assessment and what you thought the assessment should be.

All this was miles too cumbersome for Gerald Eve & Co. TSD had, ten years before, designed our own proposal form and got it settled by Jack Willis as being legally valid. Jack advised us to put the grounds of appeal as:

'1. The Valuation List is bad in law
2. The assessment in the Valuation List is incorrect, excessive and unfair and should be reduced.'

Sometimes worried VOs would telephone or write to us asking why the whole of their precious Valuation List, which they had taken years to produce, was 'bad in law' – in which case it would be null and void together with every assessment in it; quite a worrying thought.

It was a tricky question which Jack Willis had never told us how to answer.

Once when I was asked the question I said:

'Did you sign the List before depositing it with the Rating Authority (e.g. the local council)?'

'I most certainly did,' came the reply.

'Well, then, I should think you'll be all right – unless of course something else occurs to us.'

Then there was the question of to what figure we thought the assessment should be reduced. We had no idea as yet – we might well have never seen the property when we made our

Proposal, let alone valued it. So we decided to answer that question by putting 'Proposed assessment: Gross Value £1, Rateable Value £1'.

If we managed to agree an assessment that was £1 lower than that we had proposed then surely the client would not quibble.

By 31st March 1974, the last day for submitting proposals if they were to be retrospective to the date the Valuation List came into force, the full extent of the rating appeals became apparent. I was in charge of the shop rating department and we had submitted about seven thousand Proposals. The VO had twenty-eight days to object to a Proposal and if we did not withdraw our Proposal within a short period, then the Proposal became an appeal to the Local Valuation Court. (LVC). The LVC only needed to give you fourteen days' notice of a Hearing, but were usually kind enough to give you more. You could also ask for an adjournment but the Clerk would be reluctant to give you more than one.

I realised that we had an enormous task on our hands and I had better work out the best way of dealing with it. There must be a lot of streamlining and I would need many more valuers.

One of the delightful things that we did in my department occurred the day after we had been out of the office for a few days settling rating assessments with VOs. We would call in a secretary and dictate a report to the client concerning each assessment, saying what reduction, if any, in the assessment we had agreed and recommending he accept it. We might add a bit about other rents and assessments of comparable properties nearby.

If it were a young assistant dictating, he might stop from time to time to chat up the secretary or talk about last night's television or about the weekend. Indeed, I expect this was done throughout the firm (the dictating, not the chatting up), though a few people used tape recorders.

Well, it jolly well wasn't going to be done any more – not in my department, it wasn't. Such procedure was a wholly unnecessary luxury and would probably wreck the profitability of the whole exercise.

I just had two simple problems:

1. How to procure, on average, bigger reductions in seven thousand shop assessments than could any other firm of rating surveyors on earth.
2. If possible, make a bit of money out of the job as well.

For a start, I wrote to every client with a number of properties saying that we now proposed to report to them on a standard Report Form. This would obviate their having to read the whole report. I explained that we would not be doing any less work trying to reduce the assessment, but merely saving our time in writing the report and their time in reading it. I asked if they had any objections and whether they would like to alter the form in any way.

None objected; all thought it a good idea, and some made a few slight amendments to the form. It was worded somewhat as follows:

> We have inspected and valued the above property and met the Valuation Officer to negotiate the assessment.
> We have, subject to your approval, agreed the assessment as follows and recommend you accept it:
>
> Present assessment: Gross Value 1100   Rateable Value 898
> Agreed assessment: Gross Value 1000   Rateable Value 805
>
> The revised assessment will date back to 1st April 1973 and any excess rates paid will be refunded or deducted from your next rate payment.
>
> Perhaps you would kindly let us have your instructions,
>
> Yours faithfully,
>
> Gerald Eve & Co
> Chartered Surveyors

There was space for an extra paragraph or two if needed, and the letter was slightly different if we failed to negotiate a reduction in the assessment or recommended going to the Local Valuation Court. The use of these Report Forms saved thousands of manhours.

The next thing to do was to take on and train valuers in the art of getting a rating assessment reduced. We took on young lads and lasses from university or college who had got a degree in Land Economy or Estate Management. Some of the syllabuses seemed to me to be a bit academic – one Cambridge graduate did not know how to calculate the area of a triangle – but I preferred getting them straight from their studies rather than their having worked for a while in a firm whose standards were below our own, for we felt we were better at rating than any other firm in the country.

One way of making a newcomer realise he did not know it all was to give him a folding five-foot rod and ask him to measure the width of a long corridor that was about four and a half feet wide. First he might try to unfold the rod like a concertina – which didn't work and could break the rod. When I had showed him how to unfold it, he would open it to five feet and find it was too long to put at right angles across the width of the corridor. If he folded in a foot of the rod, it would not stretch the full width of the corridor. The answer, of course, was to extend the rod fully and open one of the doors.

The next thing I invented was a standardised Negotiating Sheet which the valuer had to take with him when seeing the VO. On this sheet was recorded the details of tenure, the VO's valuation in one column, ours in the next, and then that agreed. There were spaces for Rents and Assessments of Comparables and for Negotiating Points etc.

When a valuer submitted to me a Report Form in my letter book he always attached a Negotiating Sheet so that I could assess, usually within seconds, whether or not the job had been done properly and the advice was correct.

Then I decided to compile on two sides of a sheet of foolscap (we used foolscap in those days – it was a bit larger than A4) every conceivable way of getting a reduction in a rating assessment. I had been through two Rating Revaluations, in 1956 and 1963, and had negotiated hundreds of shop assessments and supervised hundreds more, so it seemed sensible to encapsulate all my experience on one sheet of paper.

Many of the negotiating points I thought up straight away but I got more by going through masses of old files. Some of my colleagues added a few as well. I then divided all these points into headings and subheadings and squeezed them all onto two sides of foolscap.

For instance, under 'Basement' I might have:

Stairs from shop: concealed/narrow/steep/spiral/at rear.
Low ceiling height, awkward shape, inherently damp, vaults too damp to use.

I then decided we would have a three-day training course in the summer for all those who would be or were engaged in shop rating. I called it a 'Shop-in'. The valuers came not only from the London Office but from all the branch offices as well. I was determined that it would be essentially a practical exercise. We hired the Seven Hills Hotel at Cobham which is now a Hilton Hotel and which was near my house in Walton-on Thames. We divided the valuers into small teams and sent them out to value the various parades of shops in the High Streets of Weybridge and Walton-on-Thames and put them in order of merit.

Back at the hotel a member of each team had to give their reasons for their ranking of each parade. Hot debates followed which were both boisterous and invaluable. At the end of all the argument someone asked what 'the answers' were. He was very disappointed when I said there weren't any answers. It was a matter of opinion formed after careful observation and reviewing all the evidence.

Another very useful exercise was to select certain shops which had a number of negotiating points both in regard to the shops themselves and their trading positions. We then got all the experienced valuers to act as VOs and see how the newcomers got on negotiating assessments with them. This was not a little unnerving for the newcomers but most instructive. At the end, one could mention the negotiating points they had missed and perhaps show how their approach to the VO could have been improved.

Altogether, it was a very useful exercise which ended with a big party in the garden of my house with lots of fun and games. Not quite the perfect ending, though, because one lad when driving away managed to run over our cocker spaniel (she recovered). Also, about an hour after they had all gone, I found one young lad still in the lavatory – clearly, he had overestimated the amount of alcohol he could drink with impunity. I remember his name but if he is reading this let me assure him that wild horses...

Some of the newcomers or 'VO see-ers' as I called them, were girls, who did just as well as the boys, but I had difficulty with my first female assistant. I think she was the first female chartered surveyor in private practice.

I believe some of the first female chartered surveyors were valuers employed by the Greater London Council, now abolished. The first lot had had to fight their way into a man's profession and they were very tough to negotiate with. The 'toughies' having paved the way, the next lot were not so tough but they were not all that easy to negotiate with.

Men for thousands of years have earned their living by doing business deals – and to do a deal you have to agree a price. After a bit of bargaining one moves, or each moves, from his original figure until they agree one. There was no such tradition amongst women. As a result, when you had heard the woman valuer's case and she had heard yours, she was liable to say that her view was unchanged; thus no agreement was reached. Only when the date of the hearing of appeal before the Local Valuation Court loomed was she forced to think that perhaps she should go some way to meet the other person's point of view and settle the appeal.

When I first took on a female assistant surveyor, many of the older men in the office had difficulty getting used to it. For instance, one of my partners would walk through the general office in my department and say:

'Good morning, Tessa'.

I would have to point out to him that he was not to say 'Good morning' only to Tessa Gibbs – even if she was very pretty. He either had to say 'Good morning' to everyone or to no one.

Then there was the Valuation Officer. When I went to see a VO I nearly always took an assistant with me. His job, apart from learning from me, was to make notes of the meeting. Then, after the meeting, he could do most of the rest of the work, being fully in the picture. On one occasion, I brought Tessa along to see a VO and introduced her. After a few minutes the VO said:

'Excuse me, Mr Eve, but I strongly object to your secretary taking down every word I say!'

I pointed out, as I should have done earlier, that she was not my secretary, that she did not know shorthand, and that she was only doing what every other assistant always does – making a few notes.

The difficulty was that you had to treat men and women equally, but that did not mean to say that you should treat them in the same way. You were asking for trouble if you did. It was no good for the morale of the department if you called in a girl to your office to give her a 'rocket' for slovenly work and she returned to the general office in a flood of tears.

Then there were little points of etiquette: when you took your female assistant out to meet the VO, who should carry the briefcase and who should enter the room first? Personally, I let my assistant carry the briefcase – it was never heavy – and I would go first into a VO's office; on the way out, or if we were going into any other room, I would open the door for her and let her through first.

My firm never publicised the reductions in rating assessments that we achieved; it might have brought in more work but it would have made the work more difficult to do; if we had adopted that practice, the Inland Revenue Valuation Office might be a shade more reluctant to give reductions that were perhaps warranted, without putting up a fight in the Court and the Tribunal.

However, there is no doubt that we saved clients an enormous amount of rates during 17 the years the Valuation List was in force, for the next new Valuation Lists did not appear till 1990. I suppose the largest saving in rates I achieved was that for Selfridges in Oxford Street. I can in this case depart from the firm's practice of not publicising reductions, because Selfridges chose to tell the Evening Standard of their rates reduction which appeared in that newspaper on 14 December 1981:

> Selfridges, the West End's largest store and the one with the
> biggest rates bill, has had its rateable value cut from
> £1,433,305 to £1,070,805. The decision will save Selfridges
> more than £420,000 on this year's figures.

If the £420,000 is taken as an average over the seventeen years (1973-90) that the Valuation List was in force, the total saving in rates amounted to over £7m. – not bad for a fee under £30,000. The rate saving in that case was over 233 times the fee.

This reinforces the old adage that the most important thing for clients is not to employ the rating surveyors who charge the lowest fees, but the firm that will get them the biggest reduction in rateable value. So those clients who thought they were saving money by putting rating work out to tender and accepting the cheapest are mistaken. By all means negotiate fees with your rating surveyor, but make sure you get the best firm – it pays you hands down.

This advice is pretty topical as there is a new Valuation List for England and Wales effective from 1 April 2005. To any firm wanting to seek a reduction in their rates I would say this:

1. Get in touch with one or more of the top firms and ask them for the names of well-known clients for whom they act for rating – preferably with similar type of properties to your own.
2. Telephone the clients and ask them how happy they are with the service they got last time when appealing against the 2000 Valuation List.
3. Only instruct a firm that is warmly recommended. It is a highly specialised job and there are a lot of cowboys out there longing to take your money.

Chapter forty-three

'I Don't Need to do This'

The City of Westminster 1 Inland Revenue Valuation Office (IRVO) had some distinguished valuers. The VO himself was TS Cane (TSC) – a man of considerable charm and intellect but of few words. I was negotiating with him the rating assessment of a shop in Regent Street and we exchanged calculations of the area of the ground floor and basement and agreed them with slight compromises. I gave him my valuation of the ground floor and basement which I had carefully compiled to total some fifteen per cent less than the assessment. TSC made a note of my valuation and then asked:

'What value have you put on the first floor, Mr Eve?'

'On the what?' I said.

'The first floor. There is a first floor of some 1200 square feet accessed by a passenger lift within the front arcade.'

'Oh!' I said.' I am afraid I never spotted that.'

It was quite plain to both of us that my valuation would exceed the assessment once the first floor showroom had been included. TSC could have pointed this out and asked me to withdraw my proposal which was under appeal. I looked up at him in obvious embarrassment and dismay. He gave me a good hard look and a glimmer of a smile. He said:

'Perhaps, Mr Eve, you would like to take another look at the property and reconsider your valuation.'

'Er... Yes I would. Thank you, Mr Cane,' and with that I hurriedly replaced the file in my briefcase and withdrew – not my appeal, but from his presence.

A true gentleman was TSC.

Some years later, at a LVC in Westminster, I was cross-examining TSC's Senior Valuer, John Pirie. I asked him one question which clearly he could not answer and I heard TSC, sitting just behind him, prompt him with the answer which John duly gave. I turned to the Chairman of the Court and said:

'Mr Chairman, my witness is being interfered with.'

TSC raised his hands as if to show he was innocent of any accusations of sexual molestation of the witness. I continued:

'Perhaps Mr Cane would be kind enough to delay giving his evidence until I have finished cross-examining Mr Pirie.'

He prompted no more.

TSC's deputy was Mr Jeffreys, a charming, very refined gentleman with whom it was a delight to negotiate. He disagreed so politely with my valuations and when he finally suggested a compromise figure, I always felt it would be churlish and indeed indecorous not to accept it.

At the end of one long morning of hard but extremely polite negotiating, Mr Jeffries said:

'You know, Mr Eve, I don't need to do this. I've got a private income.'

Years later I recalled that remark and decided to repeat it on certain sporting occasions at the appropriate, telling moment:

> on the golf course, all square and one to play, and I'm just about to drive off the eighteenth tee and I stop;
> match point against me on the tennis court. I am about to serve and I stop;
> trying a hopelessly distant roquet on the croquet lawn to save the match. I raise my mallet and stop.

In each case, I say:

'You know, I don't need to do this. I've got a private income.'

Even if my opponents do not think the remark amusing, it certainly spoils their concentration. If you lack natural ability at games, as I do, you need to be able to talk people out of a match.

I was to experience more prompting of my witness at a LVC at some steel town in Yorkshire. I was cross-examining an IRVO valuer and he turned round to his colleague and asked him what the answer was. He told him.

I tolerated this once but when he did it a second time, I protested to the Chairman. He said:

'Well, Mr Eve, to make it fair, when it's your turn to be cross-examined and you don't know the answer, you can consult *your* colleague.' A fat lot of good that would have done me to consult the junior assistant I had with me.

'Mr Chairman, this is intolerable and most improper. I suggest that before you continue further, you consult your Clerk about the correct procedure to be adopted in cross-examination.' Then I sat down.

The Chairman had a few whispered words with his clerk

'Well, Mr Eve. If the other chap knows the answer then we think that it's best if he gives it,' said this very pragmatic Yorkshireman. I wasn't having this. I said:

'Mr Chairman. Mr Smith has given evidence and I am cross-examining him on that evidence. I can only cross-examine one valuer at a time. If the other valuer wants to give evidence later then I can cross-examine him afterwards. The procedure you are permitting, after taking advice from your Clerk, is grossly improper and I will report him to the Lord Chancellor. I feel quite unable to continue my cross-examination on that basis. That therefore concludes my case, Sir.'

I sat down. There was a shocked silence. I've blown it now, I thought. Eve, I lectured myself, why can't you keep a bit calmer about these things?

After a pause, the Chairman said:

'The Court will now adjourn to inspect the appeal properties. We would be obliged if the Valuation Officer and Mr Eve would accompany us. After the inspection the Court will reconvene and give their decision.'

The appeal concerned a test case to determine the basis of assessment for a new shopping centre that was very close to the Court. We wandered round the malls, I keeping my distance. I was still inwardly fuming – and outwardly, for there was an almost visible exclusion zone around me. The three members of the Court chatted amongst themselves and, after about twenty minutes, returned to the Courtroom. I didn't think I had the slightest chance of getting a reduction in the assessments. The Chairman said:

'The decision of the Court is to reduce the assessments as follows...'

He gave sizeable reductions in both cases. I was amazed and so delighted that I decided not to report the Clerk of the Court to the Lord Chancellor after all.

I wonder if that was what the Chairman was counting on?

Chapter forty-four

'Don't Let Them Operate!

David Trustram Eve with
his father, Lord Silsoe, 1959

In a landlord and tenant case that went before the County Court I found that my cousin, David, was counsel for the tenant, whereas I was the expert witness employed by the landlord.

David Trustram Eve, now the second Lord Silsoe, succeeded to the title in 1976 after the death of his father, Arthur Malcolm Trustram Eve QC, the first Lord Silsoe. The succession was a close-run thing because he and Peter are non-identical twins, David having stolen an early lead on his brother by a mere five minutes. Malcolm was my first cousin, being the son of Sir Herbert Trustram Eve who was Gerald's brother. The row between Herbert and Gerald is mentioned in Chapter 3, but the enmity did not filter down the generations. Indeed, Herbert's sons, Malcolm and Douglas, were good friends of Gerald, and Douglas showed many kindnesses to me.

Before I tell you about the case I had with David, I think I had better explain how his father, Malcolm, succeeded in outranking his own father, a mere knight, who died in 1936. Malcolm was created a baronet in 1943 and then a baron in 1963. He took the title of 'Baron Silsoe, of Silsoe, in the County of

Bedford', after the village in which his (and my) great-grandfather settled to hold his farm tenancy for seventy-five years.

Between the War, Malcolm had an outstanding practice at the Bar and specialised in rating and compensation cases. Before that he served in Gallipoli and won an MC, and he was a brigadier in the last War. He was chairman of the War Damage Commission and the Central Land Board – in fact he was chairman of practically everything. The *Daily Telegraph* wrote in its obituary:

> **'Champion chairman'**
> He was chairman of so many enterprises and enquiries that
> he became known as 'the champion chairman of the land'.

He was First Church Commissioner for seventeen years and also First Crown Estate Commissioner. *The Times* wrote in his obituary (see Appendix 6):

> He served both Church and State with equal devotion and
> unrivalled competence. He took on great tasks in peace and
> war from which less would have shrunk, and carried them to
> fruition with cool determination and professional efficiency.

The landlord and tenant dispute in which David and I were involved concerned premises in Harley St, London which were occupied by an osteopath. At the time, I was suffering badly from a 'slipped disc'. This was obvious to the osteopath, who was kind enough to chat to me about it before the hearing started. He ended by saying what a mess orthopaedic surgeons made of backs when they operated on them. As the proceedings were about to begin, we went back to our places and as we left, his parting shot was:

'Remember. Mr Eve. Don't let them operate!'

David Eve was an extremely able and brilliant barrister – he has retired now – and amazingly self-effacing; quite the opposite of his grandfather, Herbert. Gerald Eve & Co used to employ David frequently and I was not at all looking forward to his cross-examining me.

After I had given evidence, David got up to cross-examine. Although his cross-examination was thorough, I found I got on remarkably well in the box. The reason, I soon realised, was that I could always follow his line of questioning and knew what he was going to lead up to at the end of it. Our brains seemed to work in a similar way. The result was that my evidence stood up remarkably well and the judge found in our favour. As I left the court, the osteopath, although understandably upset at losing the case, was not above coming up to me and saying:

'Remember, don't let them operate!'

The next thing I heard was that, on the advice of David, the tenants had appealed against the County Court Judge's decision. This appeal could only be made on a point of law and was to the Court of Appeal. In due course I attended the Court of Appeal to observe the proceedings. Lord Denning, Master of the Rolls, presided. This was the only occasion that I came across the great man and I was most impressed by his courtesy, charm, quickness of brain and simplicity of speech which had a slight West Country burr.

David Eve made his submissions to the judge most eloquently and succinctly on the point of law, saying that the decision by the County Court judge had been based on the wrong legal assumptions. Lord Denning got the point straight away, agreed with it, and sent the case back to the County Court for the judge to rehear on a different legal basis.

Again we went to court and there was the osteopath who straightaway asked after my back, saying that he did hope that I hadn't let the surgeons operate on me. I assured him I hadn't.

I gave evidence again – this time on the new basis – and David got up to cross-examine me again. One of the most important tests of a person's intelligence, I think, is the ability to profit from experience. On this test, David scored full marks.

He started to question me but instead of gradually constructing a line of questioning, he started to jumble the questions up and go hither and thither. I couldn't anticipate at all where his questioning might lead to and he soon had me faltering. Such effective questioning would have caused tension in anyone, but with an already tense back I found the pain excruciating. I was shifting around the witness box all over the place.

In the event, we lost the case, the tenant got what he wanted and, as I left the Court, the osteopath came up to me, now smiling, and said:

'Don't forget, Mr Eve. Don't let them operate!'

I never have.

Chapter forty five

Tone of the List

I am now about to embark on a topic which will take up the next three chapters of this book. It is one which resulted in my clients and many others resorting to the highest court in the land to obtain justice. It is a tale of dark deeds in high places – and it's all to do with the little phrase: 'tone of the list'.

When I first started to deal with rating assessments in the early 1950s, there had not been a national rating revaluation in England and Wales since 1935. The Rating and Valuation Act 1925 provided that there should be such a revaluation every five years. The one due in 1940 was postponed when war broke out in 1939. After the war the Government decided to have a revaluation in 1953 but found the job could not be done in time; so they enlisted the help of some firms of rating surveyors in private practice and postponed the revaluation till 1956.

In the meantime the Inland Revenue Valuation Office (IRVO) instructed its valuers to assess new and altered properties in accordance with the 'tone of the list'. The list referred to was the Valuation List (VL) in the possession of the Rating Authority (the local Council) and it contained details of the assessment of every property that was rated. Excluded from rates were agricultural property, churches, etc.

The IRVO instructions to their valuers had no force of law, and after you had put your case to the VO you had to accept his decision – unless the assessment was more than the current annual rental value, which was most unlikely, if only because of inflation.

To assess a property on tone of the list (tone) you had to decide what the assessment would have been if the property had been existing when all, or nearly all, the other assessments were determined. So you looked for the most comparable property and then made appropriate additions and subtractions to the assessment to allow for the differences.

The Government decided to enshrine this practice into law. Let me quote *Ryde on Rating:*

> **Statutory "tone of the list".** The statutory enactment of "tone of the list" was effected by s. 17 of the Local Government Act, 1966. That provision was repealed and is re-enacted by s. 20 of the General Rate Act, 1967.

This meant you must value the property at what it would have been assessed at if it had been subsisting during the year before the VL came into force.

If you look up 'subsist' in the dictionary you will find it means 'continue to exist'. That is what a Subsistence Allowance is for – to enable you to continue to exist – which sounds reasonable to me.

When this Act was passed, it made things much fairer because, before it, the VO could offer you a reduction on a 'take it or leave it' basis. If you did not agree with his interpretation of tone you had no legal remedy. Now you did have one and could appeal to the LVC. The VO, knowing you had this right, obviously bore it in mind when negotiating – and, of course, so did you.

This Section 20 was indeed merely to give legal force to what had been the practice of the IRVO in the past. but it seemed to do much more than that. In this connection I will tell you about *Ryde on Rating.*

Ryde is, or was in my day, the rating surveyors' bible. It probably has been for a pretty long time and I know that Gerald Eve & Co's library had a copy of every edition.

The edition to which I am referring is the thirteenth, which was updated by David Widdicombe, David Trustram Eve and Anthony Anderson, all of whom belonged to the chambers at 2 Mitre Court Buildings, in the Temple, which used to be headed by Sir Michael Rowe and included Sir John Ramsay Willis.

The part of the book dealing with tone was written by David Trustram Eve. (I had attended a talk he had given to the Rating Surveyors Association on 'Tone of the List'.) He submitted that the effect of Section 20 (which has a marginal note 'tone of the list') was that instant tone was applicable to a new Valuation List. Thus if you appealed against an assessment, technically by making a proposal to reduce it, in the VL which came into force on 1 April 1973, the assessment shouldn't be the value as at 1 April 1973, nor the value at the date of the proposal, but should be in accordance with the tone of the list.

The VL took a long time to prepare and so obviously some of the valuations were made literally years before it came into force.

As a matter of practicality, therefore, every VO had to decide on a tone date after which he could not take into account any increase in values. After all, if for instance the VO got one new shop rent which due to inflation was much higher than those previously ruling, he couldn't use it unless he increased the rents of all the other shops in the parade, and all the other parades and all the other properties in the VL whose rental values had increased due to inflation. – for obviously inflation affects every single property in the valuation list.

Certain sorts of properties such as universities were valued on a percentage of their building costs because rental evidence was insufficient or non-existent. This method was called the 'contractor's test' and for this you had to decide on a certain date to assess building costs. The date the IRVO instructed their VOs to adopt for these costs was the first quarter of 1971.

So David Trustram Eve submitted that one should enquire from each VO as to what his tone of the list date was. I tried doing this and wrote to a number of VOs asking them what their tone date was, i.e. the date when the general level of rents equated with the general level of assessments.

There was an ominous silence. I suspect the VOs consulted their Superintending Valuers (SVs) for each Region, the SVs consulted the Chief Valuer, and the Chief Valuer consulted his legal advisers. Whether the legal advisers consulted senior judges or the Lord Chancellor I know not. After a while, the replies came back. They had decided not to play ball. This is how I got on with the VO for Westminster 1:

21 March 1977

Dear Mr Jackson

Tone of the List, Westminster 1

I should be grateful if you would be kind enough to let me know as soon as possible what you consider to be the 'tone date' in your Valuation Area.
By 'tone date' I mean the date when the general level of rents entered into equals the present general level of assessment in the Valuation List.
It would, I am sure, be helpful to all concerned if I could have this information before the Local Valuation Court hearing on 5 April 1977.

Yours sincerely, Hilary M Eve

---

1 April 1977

Dear Mr Eve,

Tone of the List, Westminster 1

Thank you for your letter dated 21 March 1977. I regret I am unable to answer your letter in the way you want. However I am able to say the Valuation List was prepared in accordance with statutory requirements.

Yours sincerely, G C Jackson, DV/VO Westminster 1.

14 April 1977

Dear Mr Jackson,

Thank you for your letter of the 1 April 1977 which I am afraid is not really very helpful. I should be grateful if you would let me know whether your inability to disclose the 'tone date' is due to lack of knowledge of it or due to a policy not to disclose it.

Yours sincerely, Hilary M Eve

———————————

1 June 1977

Dear Mr Jackson,

I wonder if you received a copy of my letter of 14 April. In case you did not, I enclose a copy. Perhaps I might hear from you on this fairly soon.

Yours sincerely, Hilary M Eve

———————————

16 June 1977

Dear Mr Eve,

Thank you for your letter dated 1 June. I humbly apologise for the delay but I had completely overlooked your earlier letter. However I regret that I have nothing to add to my original comments.

Yours sincerely, GC Jackson, DV/VO Westminster 1

———————————

20 June 1977

Dear Mr Jackson,

I have written to ask you three questions:
1. What is the 'tone date'?
2. Do you know the 'tone date'?
3. If you do know the 'tone date' is it your policy not to disclose it?

> You have deliberately avoided answering all three questions.
> Inevitably, this leads me to a further question – what are you
> trying to conceal and why? You can hardly expect me to leave
> the matter as it now is. Is it really too much to expect a
> straight answer to at least one of these questions?
>
> Yours sincerely, Hilary M Eve

After that...silence. All right then, I thought; if you're not going to cooperate, I'll have to find out the tone date myself and I'll start with the City of Westminster. I got people in my firm to dig out any properties rented between 1968 and 1973 and to express their rents as a percentage of their Gross Values for rating. I also got information on this from several other firms of rating surveyors.

There were not as many properties as I would have liked because residential properties were not generally let at open market rents, due to restrictive legislation. (The rent was to be determined as if demand equalled supply.) Also, the rent for the whole property frequently did not coincide with the various parts which were separately occupied and therefore separately assessed. I called this schedule of 182 rents in 55 different streets a 'tonogram'.

After much work, I produced what I called a 'tonograph' which was a bar graph showing rents in the City of Westminster for each year as a percentage of their assessments. This satisfied me that the tone date was 1970.

I then went to the LVC to appeal against the basic of assessment of shops in Brompton Road, east of Harrods, some of which are in the City of Westminster. I explained to the Court the provisions of Section 20. I had given the Clerk of the Court notice of the passages I would quote from *Ryde* so that he was able to mark the book at the appropriate places. Thus the Chairman was able to follow the passages as I quoted them. I also provided copies of the relevant extracts so that both the other members of the Court, and the Clerk, could follow them, too.

Tonograph for City of Westminster

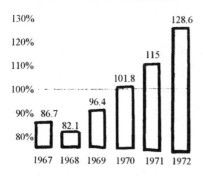

This 'tonograph' for the City of Westminster 1973 Valuation List comprises an analysis of 182 rents (in terms of Gross Value) in 55 different streets as a percentage of their assessments. The 'tone date', i.e. the date when the general level of rents entered into equals the general level of assessment in the Valuation List, is the year that coincides with the 100% line, which in this case is seen as early 1970.

Author's tonograph for the City of Westminster's Valuation List 1973

After that, I produced my tonogram and then my tonograph, explaining that since they showed the tone date to be 1970 it was the rents passing during that year to which they should have regard. The Court said they would reserve their decision and, on receiving it a while later, I found they had reduced the assessment to 1970 values.

Armed with this victory, I called a meeting of all the firms of surveyors acting for shops in the West End. There were about twenty-three surveyors who attended the meeting and seated themselves round the mahogany dining table in the Conference Room at Gerald Eve & Co.

It was generally agreed between us that we had been advised that, as a matter of law, rating assessments should accord to the tone date – after all, it said so in *Ryde*, the rating bible. When I say 'generally agreed', I mean all except one valuer who opined that the value should be as at 1 April 1973. He was shouted down. How on earth, we all asked, could a VO value everything as at 1 April 1973 when he had to start valuing years before that date? For instance, there were 8000 shops and 88,000 dwellings to value in the City of Westminster – a grand total for the whole List of some 135,000 entries covering over 6700 pages. Then the VO had to have the whole VL typed, checked and sent to the Rating Authority by December 1972.

What was he meant to do in June 1970 when he started valuing shops armed with Rent Returns from the first half of 1970?

Look in a crystal ball and hope he got it right? It was unthinkable; if he added on, say, four and a half per cent each year for inflation – the average for the last five years – and he got it wrong, the whole VL would be wrong. He was not to know that in the three years leading up to April 1973, inflation would be twenty-eight per cent. In any case, what was the point of adding on for inflation when, first, it would be a complete guess – and second, it affected every property equally? To that, the sole dissentient had no answer.

The meeting agreed that I would represent them in negotiations with the VO of Westminster 1 concerning the basis of assessment for the whole of Bond Street, Regent Street and Oxford Street. The representatives of the Association for each of these three streets similarly agreed. It was also agreed that I would represent them in negotiations concerning the north side of Oxford Street as it came within Westminster 2 Valuation Area and therefore had a different VO. I celebrated being given this job to do by giving all those present a glass of sherry.

I had taken on quite a job but the surveyors concerned seemed quite happy for me to fight their battles for them. Also, Gerald Eve & Co acted for far more department stores and shops than any other firm. In Oxford Street alone we acted for Selfridges, Debenhams, C & A and two large M & S stores, as well as a mass of shoe shops.

Shortly after the meeting, I went to see John Pirie of Westminster 1 and went through with him the whole gamut of explaining Section 20, quoting *Ryde,* and sporting my tonogram and tonograph. He was duly impressed. Not only that, but I took this occasion to wheel out my new invention the 'rentograph'. Although I proudly claim to have invented it myself, it was, like nearly all my ideas, vastly improved by other members of my department.

A rentograph comprised – and perhaps still does in the firm for all I know, even though I left it over twenty years ago – a large-scale street plan of a shopping centre showing the names of all the ground floor traders. You can buy such a Goad plan for any decent shopping street. The plan is then cut in half down the middle of the street and a strip of graph paper inserted in between. The rents for Zone A, in pounds per sq ft or metre, are then shown on a bar graph imposed on the frontage of each shop, so that wider shops have wider bars.

The colours of the spectrum are used for the various bars to show the year each lease was entered into. Thus the oldest rents – 1967 – were coloured red and the most recent ones – 1972 – coloured blue. You could remember this by thinking of 'old flames' and 'new blue whiteners'. Across this bar graph was drawn a line showing the level of assessment (Zone A price) adopted by the VO for each part of the street.

It was quite amazing how the coloured rentograph made it immediately apparent what rental support there was, if any, and at what date, for the VO's basis of assessment for each of the various stretches of a street.

On one occasion, I produced a rentograph for King St, Hammersmith, West London, when negotiating with Simon Parker of the Inland Revenue. When we had made suitable adjustments to certain Zone A prices which the rentograph had shown to be excessive, Simon said:

'Your rentograph's great for a street, Hilary, but how on earth could you do one for the shops at Hammersmith Broadway?'

He produced an Ordnance Survey map which showed an inner circle of shops facing outwards, around which the traffic gyrated, and an outer circle of shops facing inwards. On the map, he had written on the shops in different coloured inks the rents in terms of Zone A. I said I would take the map back to the office and have a think about it.

By the time I got back to the office, I had cracked it. I went to our drawing office and asked them to produce some circular graph paper – to be precise: oblong sheets of paper covered with concentric circles at small, regular intervals. This was done and duplicated and I then prepared my circular rentograph by sticking bits of the O S Map in appropriate places, and drawing the bar graphs using my own colour coding to indicate the year that each rent was entered into.

When I brought my circular rentograph back to Simon, he was so amused and intrigued he nearly laughed his head off; yet it did do the trick and clearly revealed the pattern of rents throughout the centre. We made suitable amendments to the assessments.

Evidently, he must have shown off the rentograph or talked about it to his colleagues and other rating surveyors, because for quite a while afterwards whenever I met a rating surveyor he greeted me with:

'Hullo, Hilary, been making any more circular graphs lately?'

Returning to my meeting with John Pirie, the rentograph for Bond Street revealed that the VO's basis of assessment was clearly excessive in some stretches and consequently John agreed to reduce certain Zone A prices. He reduced the whole of Old Bond Street by between 22% and 30%, and most of the parades in New Bond Street but by lesser amounts.

When I reported the reductions to the Bond Street Association they were delighted.

The next step was to meet Pirie again to agree the Zone A prices for the south side of Oxford Street. Again I prepared a rentograph – there was quite a lot of rental evidence in the street because Oxford Street shops change hands quite often, for only the most efficient traders find the enormous rents they have to pay are economic. Retailers love the street because, they say, it has 'six Saturdays a week'. The rentograph was the longest we had ever prepared because it stretched all the way from number one, next to the Centrepoint office block, to Marble Arch in the 500s.

John Pirie and I looked at the two colours, green and yellow representing the 1970–1 rents and found that everything on the south side of the street east of its junction with New Bond

Street was correctly assessed, but everything to the west of it was assessed too high. We reduced the Zone A prices there by 3% up to 16%.

I reported back to the various surveyors and they happily went and agreed the individual assessments of their clients' shops in accordance with the agreed Zone prices.

For my part, I agreed all my clients' assessments, as did my partner, Michael Hopper, in regard to C & A and the M & S store known as the Pantheon. The next step was to go and see the VO for the north side of the street – always the more valuable side as it contained nearly all the department stores.

I went to see the Deputy VO, Mr Cropper, a very experienced valuer, and again showed him the rentograph and the bases of assessment agreed for the south side of the street.

He accepted the 1970–1 basis from which it was at once apparent that the assessments on the whole of the north side were too high – except for one small parade next to Centre Point where there was no rental evidence at all.

It was clear from an analysis of the rental evidence that, whereas the rents of 1970 and 1971 were identical to each other, there had been a substantial rise in rents in 1972. Starting, I decided for tactical reasons, at Marble Arch, we reduced all the Zone A prices to accord with 1970–1 values, until we came to the parade next to Centre Point which I had decided to keep till last. Mr Cropper was adamant that he would not give a reduction because there were no rents to show the Zone A price was excessive. I said:

'Look, Mr Cropper, I know I can't prove that the assessments here are excessive, but equally you can't prove they're not. But on the law of averages, since the Zone A price for every other pitch is agreed to be excessive, surely it is highly probable that this pitch is also excessive.'

I thought I detected the slightest of signs that he might just possibly be weakening in his stance, so at once fired my last remaining bit of ammunition.

'In any case, Mr Cropper, standing back from it all, how can I sell this whole package we have agreed this morning if I have to tell this little Centrepoint lot that they are the only traders on the whole of this side of Oxford Street all the way up to Marble Arch who aren't going to pay less rates? It might start an argument – one which could jeopardise the whole agreement which we have managed to thrash out after much hard negotiating.'

'Well, all right then. Just this once,' said Mr Cropper, 'for the sake of getting an overall agreement, but I'm only giving you two and a half per cent off.'

'Done!' I said. 'Now, for God's sake let's go and have a drink and some lunch.'

But it wasn't 'done' as it turned out, for when the Deputy went to see his Chief, Mr Mabey, he was not at all a happy bunny. He got on to me on the telephone and said he couldn't accept the overall agreement.

Of course, I kicked up one helluva fuss asking him what on earth was the sense in appointing a highly experienced valuer, the Deputy VO for Westminster 2 – an area which contained the most valuable shopping street in the country, if not the world – and then not backing him when, in possession of all the rents, he had hammered out an agreement with me which was acceptable to all.

But my expostulations were of no avail. He was adamant and asked me to meet him again. This I did – I had no option – but, when I met him, there was no sign of Mr Cropper. He was hiding somewhere, or perhaps under close arrest, or even, perhaps, posted to Timbuktoo. What went on between the VO and his Deputy we shall never know, but Mr Cropper must have felt rather humiliated by not being supported by his boss in such a large and important agreement.

At the meeting with Dennis Mabey I went through my usual spiel quoting Section 20 and *Ryde* and plying him with my tonograms, tonographs and rentographs. I also stressed my most powerful negotiating point – the fact that I had agreed with Mr Pirie the Zone A price for every parade on the opposite side of the street. What was more, all these individual assessments had now been finally settled and Forms of Agreement (Forms CV/R 104) had been signed by all three parties.

In the event, I had to accept slightly smaller reductions in some parades, but by the end of the meeting I had still obtained a reduced basis of assessment for every single shop and store along the whole of the north side of Oxford Street, peaking with a reduction of 30% at Marble Arch.

To put it mildly, I was pretty pleased.

My successful efforts in Oxford Street, were greeted with acclamation all round – by my partners and by other rating surveyors. When the agreements had been converted into reduced rate demands and substantial rates refunds, our clients were very happy too; for even a small percentage reduction in assessments such as those on Selfridges and M & S produced substantial savings in rates. At the time we did not know, but these savings would last for seventeen years – until the new Valuation Lists came into force on 1 April 1990; so, as it happened, the total savings in rate payments were more substantial than we could ever envisage at the time:

The last stop in this exercise was to go for a triple whammy by getting a reduced basis of assessment for Regent Street. And thereby hangs quite a long tale – and a rum one. It'll take two more chapters to tell.

Chapter forty-six

'They're Gunning for You!'

With the help of my department, I got out a rentograph for the whole of Regent Street. We found the results truly amazing. The whole of the street from end to end was assessed at some forty per cent above 1970/1 rental levels. It was unbelievable, but, whatever way you looked at the figures, it was true.

I met John Pirie and said, as I laid out the rentograph for the whole street:

'You're not going to like this, John.'

He looked at it for a while and then said:

'Good Lord! Crumbs! I suppose, old boy, you'll be wanting nearly a third off the whole bally street.'

'Not if you can tell me a good reason not to,' I said.

'Well, Hilary, I can think of three reasons right away:

'One: a third off the lot is a bit much, old boy.

'Two: The old man [the VO, Mr Jackson] wouldn't like it one weeny bit.

'And three: the rating authority [Westminster City Council] would have fifty fits,' he said, in that rather casual drawl of his which concealed a very shrewd brain. John was a delightful man whose principal enjoyment in life was designing point-to-point courses.

'I'll tell you what I'll do, Hilary. I'll get on the blower to Westminster Council. They've appointed a valuer to look after some of their rating interests and they can send him along here. Then the three of us can try to work something out. OK?'

'OK,' I said. 'Thanks,' and John went out shaking his head, his parting remark being:

'This really is a bit dodgy, old boy. I think I can smell trouble.'

How right he was – but fortunately I didn't know that at the time.

A few days later the three of us met at my office. For the benefit of the Council's valuer, I went though the whole gamut again: Sec. 20, *Ryde*, tonogram, tonograph, LVC decision on Brompton Road, rentographs of Bond Street and Oxford street and all the bases of assessment we had agreed with Pirie and then Mabey.

Then I showed him the Regent Street rentograph and went through each stretch showing him the VO's Zone A price compared with the 1970–1 rental evidence. When I had finished, I said:

'Now you see why I need nearly a third off every assessment in Regent Street. I know I'm asking a helluva lot, what with Austin Reed, Dickins & Jones, Liberty's, Hamleys, Aquascutum and the rest, but as you can see, anything less would be unfair to the shopkeepers of Regent Street. If you can find any fault in my reasoning I would be delighted to come up in my Zone A prices.'

'To be honest, I can't fault it,' said the Council's valuer, 'but don't quote me.' (I didn't quote him then, but perhaps it's all right to do so now, twenty-eight years later.)

Then he and John Pirie said they'd both report back to their chiefs and then come back to us. They did – with bad news. The Council would not accept my basis of valuation – it would mean a third off the whole of Regent Street – an enormous loss of rateable value. It was politically unacceptable to accept the loss of millions of pounds of rateable value without putting up a fight. And what's more, they expected the VO to defend his Valuation List in Court. Surely he could not admit that he could be so wrong in his valuations.

Mr Jackson was in a difficult position; he had not made the valuations – Mr Mabey had, and the VO then (at least until August 1973) was a Mr PJ Borrett. Apart from the Council, we can only conjecture what pressure was brought to bear from above on poor old 'Jacko'. He told me there could be no deal; he had to defend the List.

We went to the Local Valuation Court at Caxton Hall on 20 October 1976, the test cases being K Shoes, at 203 Regent Street, and Saxone at 241 – both were shoe shops about half way up the west side of the street.

I cannot recall any of the proceedings but I have before me .a copy of the Court's decision dated 29 November 1976. The VO contended that Sec 20 ('tone') only applied to new shops occupied after the VL was made. The Court accepted this.

> 'The Court concludes that the Valuation Officer may produce
> rental evidence for 1972/3 to support the trend which has
> been taken into account by him in arriving at his valuation.'

That was the legal aspect, What did the Court think about the valuations?

The Chairman of the Court was a chartered surveyor who took a great interest in the rents. He dismissed some, and analysed others differently from either party. He was wrong in law to do that using his own expertise, but should simply have considered the evidence given by the valuers.

My Zone A price of £175 was based on 1970/1 values, so that was too low. The VO's Zone A price of £250 was not supported by rental evidence, so that was too high. But, like Goldilocks, he found a Zone A price of £215 was just right.

The Court allowed the appeals and reduced the assessment of K Shoes from Gross Value (GV) £33,550 to £28,800 and Saxone from GV £15,350 to GV 12,500. The VO appealed against the decision. So did we. So did the Council.

It is amazing how long lawyers take to get things done. Just look at this timetable:

December 1976	We and the VO each appeal to the Lands Tribunal.
February 1977	Westminster Council responds to the appeal.
August 1978	We submit our Statement of Case.
July 1979	Reply of VO.
August 1979	Reply of Council.
30 June 1980	Start of Hearing at the Lands Tribunal
17 July 1980	End of Hearing at the Lands Tribunal
13 October 1980	Decision of the Lands Tribunal.

But I am jumping ahead nearly four years.

Some considerable time after the LVC decision, I met John Pirie to agree facts before the Lands Tribunal hearing. I happened to meet Mr Jackson, in the corridor. He greeted me and said:

'Mr Eve, I have been having a chat with John in regard to the tone date for my Valuation List. I will concede it is the first quarter of 1971, because it is that date that Head Office have instructed us to take when working out building costs for rating valuations on the 'contractor's basis'.

We agreed therefore that when I sent to the Tribunal, as part of our Statement of Case, my agreed rentograph for Oxford Street, I would omit all rents after 1971, as they were irrelevant to a 1971 tone date.

About seven months before the Lands Tribunal hearing, I was again at Mr Jackson's office, seeing another valuer on another matter, when he called me into his office. He looked either way down the passage and then closed the door firmly.

'Hilary, I have been taken off the Regent Street appeal, so will be unable to give evidence that my tone date is the first quarter of 1971. I am sorry about that but there is nothing I can do about it. Pirie has been taken off the case and my successor as VO, Mr Hardy, is taking over as from January 1980.

'As you know, I am at the age of retirement in the IRVO but I thought it was agreed that I would stay on for a while, particularly to give evidence in this case, as we are so busy. However, I have just been informed that I am going to retire now, so I will say goodbye to you.' We stood up and shook hands and, just before he opened the door for me, he paused for a moment and said in an urgent, low voice – and I will forever remember his exact words:

'Hilary, I think you should know something; they're gunning for you.' And before I could ask who, or why, he opened the door and ushered me out.

For him to say any more would perhaps have been in breach of the Official Secrets Act.

The next time I saw John Pirie – on an unrelated matter – he said to me:

'Hilary, you know this old Regent Street lark. Do you know who caused this whole bally kerfuffle?'

'No,' I said, intrigued, 'who?'

'Well, it was old Dennis Mabey, the guy who overruled poor old Cropper. He was in our Valuation Office in Borrett's time. He found that Regent Street was a bit short on rents but that quite a few were coming up for review or for renewal of the lease. So he decided to wait till the very last minute before assessing them. As soon as he got the 1972 rents he put the whole bally street on that level. Of course it was far too late to revise any other assessments in his Valuation List and makes a mockery of what you and I agreed in Bond Street and Oxford Street – and in particular what Dennis himself agreed with you for his side of Oxford Street.'

'Well,' I said, 'I think that's jolly unfair and I am going to go and see your new chief to complain about it.'

'I wish you luck, old boy, but for Christ's sake don't tell him what I've just told you.'

'Course not! My lips are sealed, and in any case as they used to say in the War: " Careless talk costs lives".'

But the new VO, Mr D B Hardy, was not interested. He was a 'Company Man' who had been drafted in to fight a battle, and fight it he would – just like the 'Company Lawyers' told him to. He was a youngish, thin, very intelligent man with whom, sadly, I was unable to strike up any sort of rapport. He seemed to me to be earnest, intense, serious-minded, unsophisticated and quite incapable of small talk. If he had a sense of humour, it was certainly quite different from mine. I couldn't even get a smile out of him.

It was a bit like Monty taking over from Auckinleck in the Western Desert at El Alamein. Any plans for further retreats – such as those in Oxford Street and Bond Street, rather than Sidi Barrani – were burnt. Regent Street was to be General Hardy's El Alamein. But I don't think Mr Hardy went so far as to turf out all the Inland Revenue valuers and clerks from their offices to do PT and go on runs.

After we had appealed against the LVC decision we instructed counsel – and who better to choose than David Trustram Eve, now Lord Silsoe, his father having died on 3 December 1976 – just four days after the LVC's decision. David was building up a great reputation at the Bar. More specifically, he had lectured the Rating Surveyors Association on 'Tone of the List' and written the relevant chapters about it in the latest (13th) edition of *Ryde on Rating.*

David advised that we were right in law to value Regent Street on tone of the list in accordance with the tone date – not surprising really as we had been guided all along by what he had written in *Ryde.* He said he would be happy to take the case, but when the date of the hearing before the Lands Tribunal became known, David said he would have to cry off as he was otherwise engaged on some long, unending Public Local Inquiry on a planning matter. I think it concerned an atomic power station up north. David was renowned for evolving magnificent filing systems so that even though the Inquiry had been going on for months he was able to say within seconds who said what about what and when.

I was bitterly disappointed at not getting David, for there is no doubt he would have been by far the best person to defend his interpretation of the law as submitted by him in *Ryde*. In addition, it was quite clear to me that he had a brain second to none and had given far more thought to the matter of tone than had any other lawyer.

In the chambers below David's was Frank Layfield QC who agreed to appear for us, his junior being Anthony Anderson, one of the authors of the latest edition of *Ryde*.

I was more than delighted when I heard that Charles Walmsley was billed to hear the case at the Tribunal, as I had had a successful outing with him in *Siggs v RACS*, as I have recorded in Chapter 31. I very much admired his brain and his personality. Also, the very careful and objective way he listened to, recorded and considered all the evidence. He was always courteous, patient and enjoyed wit. His questions were few, shrewd and showed a real understanding of the case. But one member of the Lands Tribunal did not always understand the evidence and perhaps should have retired a little earlier than he did. Jack Ramsay Willis QC told me about a rating case he once took concerning a shop in Bournemouth. On the third day of the hearing, Jack said that the member interrupted him and said:

'I don't see why all the principal shopping streets in Bournemouth are assessed at Zone A. Surely some are more valuable than others.'

The Lands Tribunal Hearing took place at their premises in Hanover Square, London, which was but a stone's throw from Gerald Eve & Co.'s offices at 18/19 Savile Row. This was most convenient as each day our lot could come back to the office and discuss matters over luncheon. The hearing lasted fourteen days between 30 June and 17 July 1980 – over three and a half years after the appeals against the LVC's decision.

Frank Layfield presented the case competently but was not very pleased with me when I answered one of his questions in my examination–in–chief. I should explain that Counsel has in front of him your proof of evidence, unlike the witness who is usually not allowed to take it into the witness box. Counsel said, referring to a schedule of calculations in one of my exhibits:

'Would you say that it is robust?'

Robust? I pondered. How can a schedule be robust – by mounting it on card? I thought of Geoff Powell's joke about sending someone a 'stiff note' by mounting it on card. Then I remembered how my father always used to come down to breakfast in the morning and greet me with:

'I trust you are in robust health.'

Flustered and annoyed, I thought: why the hell can't you ask me questions which are in my proof? I said:

'I haven't the faintest idea what you are talking about.'

At lunch, I was advised by Her Majesty's Leading Counsel that if in future I did not know the answer to his question a better answer would be:

'I am not sure I understand that question. Would you be kind enough to rephrase it?'

I fancy he might have put it somewhat more strongly if at the time he had not been enjoying my hospitality at lunch.

Then came cross-examination, which I found a bit easier. That was by the VO's QC, Bill Glover, a jovial, witty chap. In the course of cross-examining me, he said:

'What did you do then?'

'What I had to do,' I said.

'I didn't ask you what you had to do. I asked you what you did do,' he said.'

'I did what I had to do…' and Bill Glover said sotto voce in his best John Wayne accent:

Cartoon appearing in GEN, concerning
the K Shoes case in Regent Street

'A man's gotta do what a man's gotta do.'

It was some minutes before all the laughter subsided.

At the end of one day, I remember Bill saying to me facetiously:

'Hilary, I think the first person to find out what this case is all about will win it.'

Bill was nothing if not honest. It was fortunate that he had an outstanding Junior (older than Bill but not a silk), Alan Fletcher, who was quite brilliant.

Westminster Council also dug their oar in and employed Richard Tucker QC and George Bartlett (who today is President of the Lands Tribunal) as his junior.

Mr Hardy gave evidence very well. He was most impressive. You could see why he was brought in by Head Office. When cross-examined as to the meaning of 'tone of list' he said: 'I now understand that it refers to levels of value appearing in a valuation list' (as if he hadn't always known) and when asked to what year's rentals the tone coincided with (the tone date) in Westminster he said he had no idea, because 'I'm a new boy here'. Very convenient. A pity it was against the rules of the game for me to tell the Tribunal what Mr Jackson told me – that it was the first quarter of 1971.

The Council had got rid of the valuer to whom I had given my deposition on Regent Street with John Pirie (who was keeping well clear) and they produced Mr Hampsher who was a retired VO for Tower Hamlets. His evidence was broadly similar to the VO's.

Three months later, Mr R C Walmsley FRICS read out his decision. He recited the VO's evidence as to when the VL was prepared:

June 1970	Return forms [for rent details, etc.] issued but none sorted
January 1971	Returns sorted into streets.

April 1971	Valuation of dwelling-houses started.
Late 1971	Valuation of Bond Street undertaken.
Early 1972	Valuation of Oxford Street undertaken.
April 1972	Valuation of Regent Street started.
June 1972	Typing of the list started.
August 1972	Valuation of Regent Street completed.
December 1972	List transmitted to the rating authority.

Departing from Mr Walmsley's decision for a moment, this timetable was not in accordance with the *Revaluation 1973 – Programme of Work,* produced in evidence by the VO, as contained in the Valuation Office's Circular 36 issued on 8 October 1969. This 'set out in broad outline a programme and timetable for completion of the various tasks in local offices towards the production of the New Valuation List' and is set out below. (I have put shops in bold lettering and the dates shown are the four quarters of each year.)

Rent Returns:

Issue & Sifting of Dwelling House Rent Returns	4/69–2/70
Analysis of Dwelling House Rent Returns	1/70–3/70
Issue of Shops & Misc. Rent Returns	**1/70–2/70**
Analysis of Shop & Misc. Rent Returns	**2/70–3/70**
Conversion of Surveys of Shop & Misc. to Metric	**2/69–1/70**

Precis & Valuation Summary Sheets

Dwelling Houses – Initial Entries	2/69–3/69
Abstraction of CV/R/57 Data	3/69–4/70
Shops & Misc. – Shops Initial Entries	**4/69–1/70**
Misc. Initial Entries	4/69–1/70
Shop Survey & Rent Details	**2/70–2/71**
Misc. Survey & Rent Details	2/70–2/71
Rental Analysis Details	**3/70–3/71**

Valuation – Dwelling	4/70–4/71
Shops & Misc. Hereditaments:	
Misc. – E.C.V. Classes	2/69–2/70
Shops & Other Misc.	**3/70–4/71**
Estimate of Total Rateable Values by Classes	1/71–4/71
Transfer of Revaluation Entries to Record Cards	1/72–4/72
Typing New List	2/72–4/72
Checking and Totalling the New List	4/72
Preparation of Statistical Analysis as at 31.12.72	4/72

Also given in evidence was that two years later, on 29 November 1971, the Valuation Office issued Circular 79 which stated:

> The final stages of the Revaluation – with virtual
> **completion of the valuation work certainly not later**
> **than 30 April 1972** [my bold lettering] should therefore be
> planned by Superintending Valuers with each Valuation
> Officer according to the needs of each office.

So it is clear that the completion of the Regent Street valuations in August 1972 was four months later than the absolute deadline laid down in Circular 79.

Returning to the Lands Tribunal decision, Mr Walmsley decided:

1. The 'Stop Date' (the date when no more rents could be considered) for Regent Street was late 1971.
2. At that date, Regent Street rents had been rising for the last three years.
3. To accept the VO's evidence that 'On the rental information in 1971, using reasonable expectation at 1972 in terms of a 1973 valuation list I would suggest a shade over £200 [Zone A].'
4. To adopt, therefore, a Zone A price of £215, which happened to be the same as that decided by the LVC.
5. The Gross Value of Saxone at £12,850 (LVC decided £12,500) and K Shoes at £28,150 (LVC £28,800)
6. Each side should pay their own costs.

In December 1980, all three parties appealed against the decision – appeals being allowed only on a point of law. The VO's (and the Council's) point of law, was whether under Sec 20 he ought in law to have valued the properties on rental values current on 1 April 1973. Our point of law was whether, bearing in mind Sec 20, the Tribunal came to a correct decision in the light of the evidence before him.

After giving the matter much thought and consulting my partners, we decided to accept the Tribunal's decision and withdraw our appeals if the two other parties did likewise.

No dice, so we asked Frank Layfield if he could represent us at the Court of Appeal and he agreed.

It was at this point, looking back today, that I now suspect some dark deeds in high places may have occurred. Let me explain.

The Court of Appeal decision which we were going to rely upon and cite as a precedent in our case in the same Court was the *Peachey* case in which Lord Denning, Master of the Rolls, presided. The gist of this decision was that, though it was desirable to have a VL as up-to-date as possible, it mattered not what the level of assessment was in the Valuation List so long as it was all valued with reference to the same date. Thus the assessments of all properties compared fairly with each other. The VO must refrain from taking a new rent into account when assessing a property unless that rental level can be reflected in the VL as a whole.

Quite right too. That meant, for starters, that Lord Denning would give Mr Mabey a hard rap on the knuckles for his last-minute (August 1972) valuations of Regent Street – those would be quite out of order. This decision was sterling stuff and I looked forward to Denning reiterating this principle of fairness in the Court of Appeal.

Lord Denning, sometimes known as Dissenting Denning by his legal brethren, had made himself unpopular with the Government and some of the judges – particularly the Lords of Appeal in Ordinary (the law lords)- in giving some of his decisions. He was also unpopular with some barristers as they said that some his decisions were too creative and tended to bend the law on the grounds of equity or even humanity. (Perhaps he anticipated the Law of Human Rights which we have today.) Such decisions, they complained, made the law uncertain – which was a bad thing.

Here are the bald facts that fuel my suspicion of 'dirty work at the crossroads'.

1. Frank Layfield said he would take the case.
2. After a long delay, a date was set for the hearing by the Court of Appeal
3. The Government asked Layfield if he would preside over a Public Local Inquiry concerning a proposed nuclear power station in Suffolk.
4. Layfield accepted.
5. Instead of withdrawing from our case, he asked the Court of Appeal to postpone the Hearing.
6. The Hearing was postponed until Lord Denning was about to retire who therefore could not preside over the Court as Master of the Rolls.
7. Sir Patrick Browne – ex-Treasury Counsel – prepared the judgment at the Court of Appeal.
8. A week before the case was to be heard, Frank Layfield withdrew from the case, leaving his Junior Guy Roots to take the case without him.
9. Frank Layfield got a knighthood.

Having watched on television such programmes as *Yes, Prime Minister* and *Judge John Deed,* I now have a better idea of how things are arranged in the corridors of power, or, as I put it earlier, how dark deeds are done in high places. One or two questions occur to me:

> Did some civil servant ask Frank Layfield to lunch at his club?
> Did the civil servant say that the Lord Chancellor and the Government were getting a little worried about the health of Lord Denning?
> Did he say Denning had previously rejected all suggestions that he might retire?
> Did he say that Denning was, perhaps, getting just a touch senile?
> Did he suggest Denning might be unable to do justice to this important case?

Did he want to ensure that Sir Patrick Browne, *ex-Treasury Counsel*, prepared the judgment in the Court of Appeal rather than Denning?

Did he say that the Secretary of State for the Environment thought that Layfield would be the best man in England to preside over this important new Planning Inquiry concerning a proposed atomic power station?

Did he say that for these reasons it would be better for Layfield to tell his clients that he would still take the case at the Court of Appeal, *but would get it postponed. Thus Denning would be on the brink of retirement?*

And lastly, did he indicate, by giving a Monty Python 'nudge, nudge, wink, wink', that at the end of all this Layfield would get a knighthood?

Well, I couldn't possibly comment – except to say that that's how it all worked out:

Layfield took the Planning Inquiry.

Layfield stayed on our case but got it adjourned.

Denning retired two weeks after the date of the Court hearing.

Layfield cried off at the last minute when our case came before the Court of Appeal.

Sir Patrick Browne, *ex-Treasury Counsel*, did indeed prepare the judgment at the Court of Appeal.

Layfield was 'unavailable' when our case came before the House of Lords.

Lord Bridge, *ex-Treasury Counsel*, was one of five law lords to hear the case.

Finally, Layfield got his knighthood.

Well, there you go.

It may be prudent for me to stress that in no way am I suggesting that the two judges who had been Treasury Counsel exercised their powers other than judiciously. But it is my experience that Counsel who have acted frequently on behalf of the Inland Revenue in rating cases have over the years acquired a certain ingrained attitude in regard to the question of 'tone of the list' which is inevitably reflected, albeit unconsciously, in their judgments.

Chapter forty-seven

'If the Law Supposes that...'

'If the law supposes that,' said Mr. Bumble...'the law is a ass – a idiot.'

Oliver Twist – Charles Dickens (1812–1870)

The Court of Appeal is in the Law Courts in The Strand, London. The three judges who heard our appeal over four days in 1982, 25 to 28 October, were Lord Justices Stephenson and Kerr, and Sir Patrick Browne.

Sir Patrick Browne was an ex-Treasury Counsel – which meant the Treasury retained him to act for the Government. I had worked with him before in a rating appeal to the Lands Tribunal (Chapter 30 – Revenue Robin), when he advised against going to the Court of Appeal even though there was an error on the face of the Tribunal's decision.

He had done rating cases for ages and was imbued with the tradition that 'tone of the list' had no legal force; it was purely an administrative instruction to VOs during the period between national rating revaluations – which was true until Sec 20 was passed (originally Sec 17 of the Local Government Act 1966). He still did not accept that values should accord to the tone of the list, the expression 'valuation according to tone of list' being merely a marginal note in the Act which, it had been held, does not form part of the Act and therefore has no force of law.

No one seems to have spotted in any of the Court proceedings on this matter – I have not until now – that Sec 30 (2) of the Local Government, Planning and Land Act 1980 starts with this passage:

> *(2) In section 20(1) of the 1967 Act (valuation according to tone of list)…*

This is no marginal note but is in the body of the text and surely, to put it at its lowest, indicates what Parliament, the Parliamentary draftsman and others thought had been enacted. To confirm this, read what Mr R H S Crossman, Minister of Housing and Local Government, said in the debate on the Second Reading of the Local Government Bill on 14 July 1966 (Hansard, col.1266):

> It has been the practice, since Valuation Officers took over the job 16 years ago, to value new property in what is called the "tone of the list". The Bill provides an overdue statutory basis for this fair and sensible practice of valuers.
> It lays down that the value must be fixed as though the valuation was being made when the list was being prepared: that is, in 1962. [That List came into force on 1 April 1963]

Unfortunately, it was not Mr Crossman who prepared and delivered the judgment in our appeal. It was Sir Patrick Browne and his compatriots, who decided that Parliament enacted that Regent Street should not be valued *when the List was being prepared* but at values existing *at the date the valuation came into force* – in this case, 1 April 1973.

It is interesting seeing the difference between a judge with a jury and a judge without one. When I sat on juries, which I have done a number of times, the judge was invariably most courteous to us and was anxious not to alienate us in any way; thus we would be purely logical when coming to our decision. (Fat hope! – but I must not disclose the secrets of the jury room.)

However, once you get judges without a jury, they make no effort to hide their irritation and are sometimes downright rude – albeit in a cutting or sarcastic sort of way. They are at their worst after lunch. A post-prandial Browne was a real grump and vented his anger on me for inventing the words 'tonogram', 'tonograph' and 'tonometry', each of which I had defined before using. I remember his going scarlet in the face and spluttering:

'This tono...BUNGLE!'

This was not the calm, dispassionate judgment I had hoped for from the second highest court in the land.

Browne read out his decision, with which the other two judges agreed, on 9 December 1982. He dismissed my evidence as to tone date by saying that I had produced in support only 51rents out of 135,000 assessments in the VL. I had in fact produced 183 rents in 55 different streets, and 88,000 properties were dwellings with virtually no market rentals. Now who is the bungler? In any case, if the list were meant to be uniformly valued at the same date – whatever that was – 183 rents would be a good indication of tone date. It would be rather like taking the temperature of a pool (all the water being at the same temperature) in 183 different places. There may be thousands of other places in the pool where one might take the temperature but you would get the same result. In any case, mine was the *only evidence* as to tone date.

Amazingly, concerning our submission that tone of the list is the date rents coincide with assessments, Browne said,

'I can't accept that for a moment.'

What else could it be?

Browne rejected the interpretation of the law as variously interpreted by Mr Walmsley in the Lands Tribunal, by our counsel, and by David Silsoe in *Ryde on Rating*. Everything must be valued at 1April 1973 values and, as the proposals to reduce the assessments of the two shops forming the test cases were made in August and September 1973, the VO can use rents up to those two dates to see if he got the assessments right when he did his valuation the year before. It would follow from this that if the shop next door made a proposal at a later date, say March 1974, even later rents could be taken into account, justifying a higher basis of assessment than his neighbour. That is absurd.

Browne found in favour of the VO and the City of Westminster and ordered us to pay half the other side's costs in the Court of Appeal and the Lands Tribunal. The original assessments were restored. The judgment ended thus:

> In conclusion, we should like to express our great debt to Mr Roots. He was bereft of both his leaders, but no one could have presented the case for the ratepayers more ably or attractively, and his clients did not suffer the slightest disadvantage from the absence of leading counsel.

I fear I thought differently, for, after the hearing, I made a list of seventeen points where I thought counsel either went astray or omitted something.

I had some sympathy for our counsel for, not being a silk, he had to sit in the second row to make his submissions to the Court, while the silks for the VO and the City of Westminster lorded it in the front row.

The lawyers went to David Widdicombe QC, another co-author of *Ryde on Rating*. He agreed to take the case so an appeal was made to the House of Lords. The appeal was heard on the 4th and 5th of October 1983 by Lord Keith of Kinkel, Lord Fraser of Tullybelton (do they make up these place names?), Lord Scarman, Lord Bridge of Harwich and Lord Templeman of White Lackington. The other parties had their usual counsel and we were represented by Widdicombe with Guy Roots as his junior.

When our appeal went to the House of Lords, that is exactly where it was held – at least it was in our case. The five judges were dressed in lounge suits, appropriately for they lounged bareheaded on the red leather front benches furthest from the Woolsack. Two sat one side and three on the other. In the gangway stood counsel in his gown wearing a very heavy, full-bottomed wig which I am told is very hot to wear. Some way behind him sat junior counsel who could not see the judges and was so far behind his leader that he was unable to prompt him or consult with him – which in this case proved absolutely disastrous. The rest of us sat on the backbenches and had a good view of the proceedings.

It seems to me that the judge who prepares the decision is always the grumpiest one. This was undoubtedly Lord Templeman, who frequently became impatient and irritated and went very red in the face. I had already told Widdicombe that Browne in the Court of Appeal didn't like my new words such as 'tonogram', so he scrupulously avoided using them. All to no avail, for my new words were mentioned in the text of the Court of Appeal's decision which was before the judges. When Templeman came across them it sent him into a real bate – he could hardly spit out the word 'jargon'.

This coming from a gentleman of the law was pretty rich, what with its *prima facie, rebus sic stantibus, a fortiori* and the like. Templeman himself used in his decision the phrase *obiter dicta*, but that did not stop him from giving poor old Charlie Walmsley a rap on the knuckles for 'falling victim to alternative jargon' by inventing the term *stop date* in his Lands Tribunal decision. At least *stop date* is English and entirely self-explanatory in its context.

David Widdicombe QC had a reputation for being a highly competent counsel, but unfortunately in this case he was woefully inadequate. I can only put it down two things:

First, to the fact that he must have had insufficient time to read all the documents in the case – and there were a lot of them

– to take them in, to understand them fully and in depth so as to cope with cross-examination, and to decide how properly to present the case.

Second, he had very recently returned from the Far East and must have been severely jet-lagged.

There is a very big difference between counsel appearing before the Lands Tribunal, on the one hand, and before the Court of Appeal and the House of Lords on the other. In the former, it is the witnesses who are cross-examined, but in the latter it is the barrister himself. Thus in those two Courts the preparation of the case must be much more thorough; one needs, like any experienced expert witness has learnt, to soak up the background to the case.

Widdicombe evidently had not done so, even though I had given him the seventeen points that his junior failed to make in the earlier Court, for in cross-examination he gave all the wrong answers. It was unbelievably frustrating for me to have to sit there listening to all these answers being given without being able to prompt counsel or, better still, answer them myself. Guy Roots, his junior, must have had similar feelings.

As Templeman said later in his decision:

> 'The appellants' counsel was quite unable to provide any
> convincing answer to the questions posed by my noble and
> learned friend, Lord Bridge...'

When we adjourned, whilst waiting for Widdicombe to appear, one of the Inland Revenue solicitors said to me:

'I think you could have done a lot better yourself, Hilary.' I certainly could.

When Widdicombe, shaken by Nigel Bridge's withering, clinical cross -examination, joined us, he said something like:

'If you want things done the way you want, Hilary, next time you'd better go and become a barrister and take the case yourself.'

Guy Roots, having a far, far better understanding of the case than his leader, and having digested the seventeen points I had made concerning his previous performance, took what I am told is the very rare step of what I think may be termed in the profession (excuse the jargon) 'following on'.

When the hearing resumed, Guy Roots indicated to his lordships that he would like to 'follow' his leader. What he said was good, sound stuff but it fell on deaf ears. Their lordships had listened to Widdicombe, Nigel Bridge had torn him to shreds and the party was over. They had made up their minds and Guy Roots might just as well have saved his breath. They didn't listen and they didn't even want to hear what the opposition had to say. They felt that Widdicombe had made no case to answer.

On 3 November, Lord Templeman of Temper Temper, the House of Lords grump, read his decision with which the other four judges concurred. He found Sir Patrick Browne's judgment 'impeccable' which was surprising since Browne had said 'Mr Eve seemed to base his opinion [on tone date] on 51 hereditaments' whereas I had based it on 183 hereditaments in 55 different streets. Lord T of TT also said that 'the thoroughly fallacious reasoning of the appellants followed a tortuous path.'

His most remarkable statement though, and in my view the biggest compliment paid to any profession since the beginning of time, was this:

> 'it does not seem to me that the Westminster VO or any other trained valuer would have any great difficulty in estimating the rents obtainable on 1st April 1973...'

notwithstanding that it was accepted that some of the estimates were made some two years in advance of this date, based on Rent Returns received in June 1970. Only a lawyer could say something quite so fanciful. To quote Mr Bumble: 'If the law supposes that, the law is a ass – a idiot.'

What makes the statement even more extraordinary is that it flies in the face of the evidence given in the Lands Tribunal by the Westminster VO himself, who said (see point 3 of the Lands Tribunal decision on p. 386):

> 'On the rental information in 1971, using reasonable expectation at 1972 in terms of a 1973 valuation list I would suggest a shade over £200 [price for Zone A]'

In fact, it turned out that April 1973 rental values were at least £250. So much for Lord Templeman's crystal ball.

In The Estates Gazette of 4 February 1984, a letter by P M Reed FRICS of King & Co, a well-known firm of rating surveyors, summed it up:

> There is simply no way that a decision in the House of Lords can make viable the crystal ball method of "valuing forward". Any valuer working in the early 1970s who had attempted to "value forward" for *any* purpose would subsequently have been proved to be wrong, for many reasons, including the level of rents, the state of the economy and the snowballing effect of inflation.

I was minded of Fiedler's Forecasting Rules:

(1) *The first law of forecasting:* Forecasting is very difficult, especially if it's about the future

(2) *For this reason:* He who lives by the crystal ball soon learns to eat ground glass.

Early in 2004 a leading building society predicted house prices would go up by eight per cent that year, whilst a large firm of merchant bankers said house prices might fall by twenty per cent in 2004 and thirty per cent in 2005. I assume that both firms have research departments and have taken expert advice from trained valuers. One thing is certain: one of these predictions will be wildly out – and probably both of them.

According to the Halifax, in the year 2003 house prices (which are broadly related to rents) went up in Brighouse, west Yorkshire, by sixty-five per cent. In contrast, the Land Registry reported that house prices that year went down in Camden and

Islington by 5 per cent. Those are the variations in one year, let alone two or three.

The difficulty of any VO correctly predicting future values was accepted by Mr Osborn of the Department of the Environment when I went to see him with one of my partners, Alan Duncan, on 12 October 1978 – a date which was after the Local Valuation Court's decision on Regent Street but before that of the Lands Tribunal. We suggested a statutory 'antecedent date' – a date well before the Valuation List came into force by reference to which everything would be valued.

Mr Osborn wrote to us on 6 November 1978 to say that 'this idea is under active consideration for inclusion in future legislation, and it was very useful for us to have your comments on this'. This was enacted two years later in Section 30 (1) of the Local Government, Planning and Land Act 1980 which enabled the Secretary of State for the Environment to set the antecedent date – later set for two years before the new VL came into force. Thus the new April 1995 VL was based on April 1993 rental values. Perhaps you have got me to thank for that.

I thought this whole 'foresight saga', as I dubbed it, was summed up rather neatly in the *Chartered Surveyor Weekly* when reporting the *K Shoes* case:

> At the end of the day the Crown will always have its way, for if it loses a case, it will appeal, and if it loses on appeal, it will legislate. In the present case, the Crown appealed against the original decision of the Lands Tribunal in favour of the ratepayers; and, although it won in both the Court of Appeal and the House of Lords, it has also legislated. A real belt and braces job.
>
> Barry Denyer-Green LL.M FRICS, barrister,
> a principal lecturer in law

I feel sure that the only reason the House of Lords felt able to come to such an extraordinary decision was that they knew that, due to the passing of the 1980 Act three years earlier, no VO would actually have to do crystal ball

gazing to achieve the accurate predictions of which their Lordships in their foolishness (I nearly wrote 'wisdom') thought him capable.

Shortly after the House of Lords decision, I drafted a letter to Lord Denning about it (see Appendix 8), but something inside me stopped me from sending it. To this day, I cannot put my finger on what it was, because I feel sure he would have sympathised with me.

I contacted Nigel Bridge, with whom I had been on a number of cases, and asked to come and see him. He readily agreed. On arrival and after security checks I was taken up in the lift by some uniformed gent and led along lavishly decorated red-carpeted corridors to Nigel Bridge's room. I drank china tea with him from a House of Lords cup on a House of Lords saucer, and took biscuits from a House of Lords plate while Nigel was on the telephone talking to the United States Speaker, organising a visit here of their Supreme Court.

Nigel always was, and I am sure still is, a very polite, gentle person; when he slowly carved you to pieces in cross-examination he always did it very gently and with a smile on his face – rather like an expert carver dealing with the Sunday joint. He started by telling me that Templeman in his draft decision was rather rude about me, and that he, Nigel, had toned it down. I think it must have been Nigel who inserted the bit that ran: *While I acknowledge the ingenuity, sincerity and experience of the surveyor who was responsible for the "tonogram"* before the bit: *I reject this method of approach entirely.*

I explained to Nigel that, whatever lawyers might think, their decision meant that Regent Street was the only street in the England and Wales assessed for rating on 1 April 1973 values. The Inland Revenue had laid down for VOs a nationwide timetable for the various stages of seeking and collating rental evidence; for valuing; for making totals; and for typing the VL.

As a result, fairness, uniformity, and, as far as possible, correctness, had largely been achieved throughout the country by all VOs doing the same things at the same time. Valuations were done in 1971 and early 1972 based generally on the rents they had collected in 1970.

Every rating surveyor knew that – and every VO did. I was aware that this applied everywhere because Gerald Eve & Co. had the largest rating practice in the country; and this was spread throughout every county in England and Wales – particularly as regards shops.

I suggested to Nigel that, as he was a member of the legislature, he might consider introducing a short bill in the House of Lords to put this right, so that 'Tone of the List' instead of being a mere side heading in Section 20 of the General Rate Act 1967 could have the force of law – which side headings do not.

Nigel gave me tea but not sympathy; he said that my suggestion would be 'most improper', which it would not have been because Law Lords have done such a thing both before and since then. But it would certainly have been most embarrassing – and tantamount to an admission that their Lordships, in giving their decision, had invested the 'Westminster VO or any other trained valuer' with supernatural powers of clairvoyance and the ability accurately to predict rental values up to three years ahead.

Here is a little exercise you can undertake. To assist you, you may look at the graph reproduced here, showing the yearly percentage increase in house prices. Now imagine it is 1988 and you are the VO. You are making rating valuations of houses (which in fact were then excluded because they were subject to Council Tax instead) for the new Valuation List that

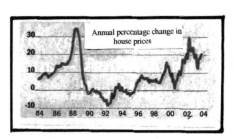

Annual percentage change in house
prices 1984–2004

is to take effect on 1 April 1990. Looking at the graph leading up to 1988, would you as a VO have 'had any great difficulty' in correctly predicting house prices two years ahead, i.e. 1990?

If in 2002 I could predict with 'no great difficulty' that house prices would go up sixty-five per cent in Brighouse in the next year (as they did), I could have borrowed some money, and bought up almost every house in town simultaneously at forty per cent above their then market value – well, wouldn't you sell your house at that price? Then I could sell them a year later and cash in on the remaining twenty-five per cent increase.

Who would lend me the money? Well, if a 'trained valuer' could predict values several years ahead I would already have built up a vast clientele because of my infallible track record.

But then, if 'any other trained valuer' could do this, we would all be buying up in Brighouse and selling up in Camden and Islington, which would affect values *then* – not in a year's time.

So, if we can predict the future, it follows, at any rate in this case, that we can alter the future. And if our knowledge of the future alters the future then our original predictions do not come about after all. What a topsy-turvy magical kingdom we trained valuers must live in! Possessing the predictive powers that the Law Lords invest us with, it is surprising that there are any of us still working. Why haven't we all retired as billionaires by now?

Let's face it; Lord Templeman was talking crystallised balls.

★★★

When reviewing the *K shoes* decision in the cold light of day, some twenty years after it was given, one might ask: if the House of Lords got it wrong, how then did one get Valuation Lists that were fair to every ratepayer throughout England and Wales? This was the most important question put by Lord Bridge to our counsel, Mr Widdicombe. The short answer is:

By all VOs doing what they were told when they were told.

Head office circulars told VOs when to send out Rent Return Forms (which had by law to be returned within 21 days), when to analyse the rents on those forms, and when to make the valuations based on those analyses.

That, to my personal knowledge, and that of my firm, is how it was done throughout the country – except in Regent Street, London.

A pity, that.

Chapter forty-eight

A Midwinter Night's Dream

'...dreaming dreams no mortal ever dared to dream before.'
The Raven. Edgar Allan Poe (1809–1849)

The House of Lords may be the highest court in the land but it is not the highest court in the European Union; there is the European Court of Justice in Strasbourg.

I remember the Duke of Westminster, reputedly the richest man in Britain, resorting to the Court to oppose the compulsory leasehold enfranchisement of some of the houses on his estates. He lost his case and I always wonder if he was still the richest man in Britain after paying all the costs.

So the European Court of Justice is not one that should be resorted to lightly and it was not one to which I suggested that the over-taxed ratepayers of Regent Street should resort. I wouldn't dream of doing so; so I thought; but I did – and here is my dream:

The scene is the European Court of Justice in Strasbourg.

The Clerk of the Court: Call Lord Templeman!
Lord Templeman enters the witness box.
French Advocate: Etes vous Seigneur Templehomme de Lackington Blanc?

Lord Templeman: I'm not going to be cross-examined in French by a ruddy Frog!

Judge (Dutch): Objection sustained.

German Advocate: Sie sind Lord Templemann von Weiss Lackington, nicht?

LT: Neither did I fight in the 2nd World War to be cross-examined by a Jerry.

Judge: Objection sustained. *Turns to German Advocate:* And when are you going to give us back our bicycles?

German Advocate: Ach, so. Ve haf vays of making you valk!

British Advocate (me): Do you call yourself Lord Templeman of White Lackington?

LT: I do.

Me: I put it to you that you are not from White Lackington at all. You have conjured up the name out of thin air to give the impression that you come from an attractive little hamlet in the heart of the country, whereas in fact you come from some wretched little suburb such as East Cheam.'

LT: Me? East Cheam? Never!

Me: You see, Lord Templeman – of wherever you are really from, you do make things up don't you?

LT: Most certainly not!

Me: Well, we'll see. I refer you to the last line on page 4 of your decision:

> "...the Valuation Officer's estimate [in 1972, for the first quarter of 1973] was accurate. So much is conceded."

I put it to you that no one conceded that the VO's estimate was 'accurate'. They said, in 1980 with the benefit of hindsight, that it was 'not excessive'. So there was no concession that the VO's estimate was 'accurate', that the VO was able to guess right, was there?'

LT: 'No.'

Me: So, in order to demonstrate that the VO could make an accurate estimate, you made up the word 'accurate', didn't you, to suit your book? You simply made it up.

LT: I wouldn't put it like that.

Me: Well, how would you put it? Where is the evidence in this case that a VO can accurately predict values?

LT: I don't know, but I still think that a VO could predict values ahead accurately

Me: But that is not your expert opinion is it? You are not a trained valuer, are you?

LT: No

Me: And there is no evidence in any of the cases quoted in this case, or indeed in any other case since the beginning of time, to support your wholly inexpert opinion?

LT: Not that I am aware of.

Me: But there *is* evidence in this very case to the contrary, Lord Templeman, isn't there? Let me refer you to this passage in the Lands Tribunal decision – it's at page 63 of Appendix Part I:

> Mr Hardy [the VO]... said: "...On the rental information in 1971, using reasonable expectation at 1972 in terms of a 1973 valuation list I would suggest a shade over £200."

So Mr Hardy's evidence was that if he was valuing Regent street (which was in fact valued somewhere between April and August 1972) he would have predicted a value by April 1973 of 'a shade over £200' as a Zone A price for Regent Street; whereas it is admitted that a Zone A price of £250 was 'not excessive'. His forecast was way out, wasn't it?

LT: So it seems

Me: It doesn't just seem, Lord Templeman, it actually was, wasn't it?

LT: Yes

Me: So the evidence in this case is that the VO cannot predict future values even one year ahead, let alone when he started valuing elsewhere in Westminster in April 1971 based on June 1970 Rent Returns. Is that not so, Lord Templeman? This evidence is exactly the opposite of what you said in your judgment isn't it? It makes a nonsense out of it.

LT: Yes, out of that particular part.

Me: I will now move on to another part of your judgment. Lord Templeman, do you think you have a reasonable command of the English language?'

LT: I like to think so.

Me: So you know the meaning of the word 'impeccable'?

LT: Yes.

Me: Do you accept the definition I have here in The Collins Concise Dictionary: 'without flaw or error'?

LT: Yes

Me: I refer you to line 8 of the last page of your decision where you described the judgment of the Court of Appeal delivered by Sir Patrick Browne as 'impeccable'. Do you see that?

LT: Yes

Me: Now look at page 17 of Sir Patrick Browne's judgment starting at line C3:

> The difficulty of discovering a "tone year" or "tone date", in the sense defined by the ratepayers (i.e. the time at which the values in the List most nearly coincided with rents in the market) is obvious. Mr Eve seems to have based his opinion on 51 hereditaments out of a total of 135,000 in Westminster (see Exhibit WCC/3).

In fact, I based my opinion of the tone date on a comparison of the rents and assessments of 183 hereditaments in 55 different *streets*. These streets ranged from The Strand in the east to near Chelsea Bridge in the south, Bayswater and Notting Hill in the west, northwards up Edgware Road and north of Regents Park.

Sir Patrick Browne mistook the list of *streets* in Exhibit WCC/3 as a list of *hereditaments*. Thus he thought I quoted, for instance, only one shop rent in New Bond Street rather than an actual 19, and only one shop in Edgware Road rather than an actual 18.

Is that a flawless judgment, Lord Templeman? One without flaw or error?

LT: No. That was a mistake but it does not alter the law.

Me: But the law says, does it not, that a Valuation List should be uniformly correct?

LT: Yes

Me: Thus if 183 assessments in 55 different streets are based on 1970 rents, then the rest of the shops in the street will be assessed in line with them; and the neighbouring streets in line with them, and so on, the whole making one interlocking jigsaw, all uniformly correct on 1970 rents, won't they?

LT: Yes

Me: Remember, all the facts in my schedule were agreed and it was the **only** evidence offered on the subject of tone date, none of it being disputed. The truth of the matter is, is it not, Lord Templeman, that the then VO valued virtually the whole of the City of Westminster during the period April 1971 to early 1972 based on Rent Returns sent out and received in 1970?

LT: Is that right?

Me: Well, it's all there in evidence in the Lands Tribunal decision. You have read the decision, haven't you?

LT: Yes

Me: And not surprisingly the general level of assessment equated to 1970 rental values. It is apparent from studying the tone of the list that, if the VO felt he had to look ahead, he had the sense to project forward horizontally – i.e. assume that rental levels would be maintained. He had no reason to allow for inflation because that affects all properties equally doesn't it?

And then, quite contrary to instructions from Head Office, some bright spark at the Valuation Office decided to hold back on Regent Street and was still valuing it as late as August

1972 **using the latest rents**. And you now have the nerve to say that it is all right to value just Regent Street on April 1973 values because rents by then had justified it. That's right isn't it?

LT: Yes

Me: So instead of Regent Street being assessed at £175 Zone A so that the list was uniformly correct, you think it is all right in law for the VO to single out this street and stick the basis up to £250, about 40% higher than the 1970 rental levels on which the rest of Westminster is assessed?

LT: I don't think it was like that.

Me: It bloody well was. (*Remember this is a dream and one is not always polite in dreams*). And what I think really takes the biscuit, Lord Templeman, and what really gets my goat, is that you say blithely at the end of your decision in November 1983, 'the ratepayer has no cause for complaint' if his premises are valued on a 1st April 1973 rental value because 'assessments which are too low can be increased on the initiative of the valuation officer or the rating authority'. In reality this means that all the assessments based on 1970 values, some 135,000 in this case, should be increased if everyone is going to pay his fair share of rates, doesn't it?

LT: If what you say is correct, yes.

Me: And did you know, Lord Templeman, that as of now, November 1983, the date of your decision, no steps have been taken to do this – not even have the 183 assessments which are accepted to be on 1970 rental values been increased.

LT: No I didn't.

Me: It's an absolute scandal. Are you not ashamed of yourself to be associated with such a legal conspiracy to bend the law in such a grossly unjust fashion to suit the convenience of the Government? Don't bother to answer that. No more questions, you silly old...

I wake up shouting angrily. Unfortunately, I had woken up before putting to Lord Templeman my favourite final question I once heard put by a barrister on television:

Me: I put it to you that the whole of your evidence is a tangled tissue of lies deliberately concocted to discredit my client.

This Valuation list stayed in force until a new one became operative on 1st April 1990, so the poor shopkeepers of Regent Street had to put up with this gross injustice for seventeen years. Which was quite outrageous.

It is some consolation that, following our representations to the Government early on in the Regent Street appeals concerning the inequities thrown up by this test case, it enacted the legislation we proposed: rateable values in a Valuation List can now by law be, and indeed are, based on an antecedent date set by the Government. So the 1990 Valuation List had in law to be based on 1988 values. About time, too.

So, Lord Templeman, contrary to what you said,

> (...it does not seem to me that the Westminster valuation officer or any other trained valuer would have any great difficulty in estimating the rents obtainable on the 1st April 1973...),

the Government felt that the valuers **did** have a difficulty – and has legislated accordingly.

Thus, no more will trained valuers need your crystal balls – but I will not tell you what to do with them – not even in my dreams.

Chapter forty-nine

'Quality is Never an Accident'

'Light the blue touch-paper and retire.'

Instructions for letting off a firework

My brother Adrian, three years older than me, died in July 1982 at the age of fifty-eight. At the time, he was Joint Senior Partner of Gerald Eve & Co. I think the highlight of his professional life must have been when he chaired the dinner the firm gave to the legal profession at Skinners' Hall in 1980 to mark the firm's Golden Jubilee at which he made an excellent speech.

Author (left) with his brother, Adrian Eve, in 1980 beside their father's portrait commissioned by Gerald Eve & Co to celebrate the fiftieth anniversary of the founding of the firm.

Two years later he got cancer of the bones and of the thyroid and within three months he was dead. It was a great shock to us all.

A younger brother is not the most objective of judges of an older brother, so let me quote a passage or two from the address of John Drinkwater QC at Adrian's Memorial Service in September 1982:

> He had achieved great eminence as a very experienced, shrewd and highly individualistic expert planning witness... His knowledge of planning matters was formidable, his approach to problems analytical, meticulously attentive to detail, and often original and innovative. He had that rare talent which is the ability to distil and communicate the essence of complex cases through apt and vivid choice of words and analogy. All this combined with great energy and zest for the job in hand.

That sounds to me as if Adrian carried on the Gerald Eve tradition pretty well.

Adrian's death made me realise that I did not want to die 'in harness'. I did not live for the job and take work home every night like some of my partners did. If a job appeared on my desk, I wanted to do it better than anyone else on earth, but there were many other things that interested me. One other interest was when the partners decided to have a house magazine. I was asked to be Editor-in-Chief and was delighted to accept.

I called the magazine *GEN – Gerald Eve News –* and much enjoyed, with assistance from my colleagues, writing and commissioning articles for it, hunting out photographs and generally putting together the magazine. It first appeared in May 1977 and I continued to be Editor-in-Chief until I retired six years later. It was published monthly and every member of the firm got a copy.

The information in it was professional, social and sporting and I thought it an excellent way to let every member of the firm know what was going on and to feel part of a larger team – after all, at that time the firm comprised about 160 people spread over six different offices.

CONFIDENTIAL
FOR INTERNAL
CIRCULATION
ONLY

GERALD EVE NEWS

VOL.7
NO.7
JUNE
1983

THE STANDARD, THURSDAY, JUNE 16, 1983—

Stores battle for Lords

REGENT Street stores which have been fighting a rates battle for nearly 10 years are planning an appeal to the House of Lords.

The decision follows a Court of Appeal ruling against two shops, which are backed by a consortium of 66 other big stores.

If they win, their rates bills could be trimmed by hundreds of thousands of pounds a year.

The Lords hearing could prove an important test case for thousands of other properties across the country.

The case involves K Shoes and Saxones, owned by the

Standard Reporter

British Shoe Corporation.

Among those backing them are Liberty, Dickins and Jones, Jaeger, Hamleys, Mappin and Webb, Garrard—the Crown jewellers—Raine—the Queen's shoemakers — Austin Reed, Aquascutum and the National Westminster Bank.

Rents

Regent Street shops pay some of the highest rates in the country—around £7,000,000.

If they decide not to take the case further they face a further bill of over £3,400,000, part of the increase withheld

until the dispute is settled.

The case centres on whether, under a 1973 revaluation they should have been rated at £250 a square metre or £215 as the shops claim.

The problem arises from differences of interpretation about the timing of the valuation and whether it should have been set to take into account the property boom of the early 1970s when rents, on which rates are based, rose sharply.

Shops in Oxford Street, Bond Street and Piccadilly pay less because they were valued at a level set in 1970. Regent Street claims it is being penalised because it was revalued later.

This appeal has been set down for hearing by the House of Lords on October 3rd. David Widdicombe QC has been retained. It was he who managed to get the GLC's Supplementary Rate quashed, on behalf of Bromley Council, in the House of Lords. Hilary Eve took the Regent Street case to the Local Valuation Court and got a 13% reduction. The Lands

Tribunal confirmed the values awarded by the Local Valuation Court, but the Court of Appeal reversed the decision and, on a point of law, found in favour of the V.O. and City Council.

Let us hope the House of Lords reverses the Court of Appeal's decision. It would not be the first time.

The issue for June 1983 of GEN, Gerald Eve & Co.'s house magazine

I enjoyed this work so much that after some years it provoked one of my partners to say: 'Haven't we got rather an expensive editor?'

I thought he had a point and, after that, having guided GEN into the sort of house magazine I wanted, I did rather more delegating.

One of my other duties, on occasion, was to give a welcoming pep talk to the newcomers to the firm. This came at the very end of their induction course after which we moved into the conference room for drinks. If you want to know exactly what I said on one occasion, it's in Appendix 7 at the end of this book.

One of the excellent things about being a surveyor in Gerald Eve & Co. was the back-up you got. If you were given a new job which was rather different from usual, you could go and talk about it at the luncheon table if you were a partner or Associate. I always thought my partners were a very bright lot and the variety of experience they had was quite amazing. Nearly always you'd find that someone else had valued that sort of property in that sort of area for that sort of purpose. He would give you a tip as to how to go about it and tell you where to look for the 'dead' file.

Then there was the filing system that had been redesigned by me (see Chapter 32), where you could look under client, type of property, type of job and location.

On top of that was GEN which told you some of the professional work other people were engaged in. There was also a New Jobs List which came round every week. However, in addition to that, we had a Research Department headed by a Ph D, a marvellous library and a librarian. It was one chap's job to look through all the new legislation, not just Acts but Regulations and Statutory Instruments, to decide which partners would be interested in which documents. These were obtained, mostly from Her Majesty's Stationery Office, immediately.

Every department was supplied with all the relevant professional journals together with copies of interesting articles which might have appeared elsewhere.

Then, a few years after GEN started, along came EVEBRIEF which included our own reports of recent cases in the Courts and the Lands Tribunal, plus information on Planning decisions.

Working in Gerald Eve & Co, I got so much back-up information that when, a fortnight after I had retired, someone asked me in Bosham Sailing Club if I knew anything about leasehold enfranchisement I said: 'Not really, now I'm retired.'

Although I probably knew about a hundred times more about leasehold enfranchisement than any one else in the bar, I felt I was out of touch and out of date because I had not been at my desk for a fortnight to receive almost daily briefings.

Shortly after the death of my brother, I told the partners I would like to retire from the firm. I had in my time 'lit the blue touch-paper' and caused plenty a firework, both in the profession and in the partnership. Now it was time to retire.

Amazingly, my partners decided they would somehow manage to stagger on without me. One of the reasons for this is that every owning partner was sent off on a four months 'sabbatical'. This was invaluable as it meant that during that period the firm had to learn to do without him. In turn, he learnt that he was not indispensable, and during that period his clients were passed on to younger people. Obviously, this would smooth the process of handing over clients when he retired.

It was agreed that as from the beginning of the next financial year, 17 April 1983, I would go onto the equivalent of a 'two day week', or to be precise, work during the year for one hundred working days, after which I would retire completely.

I started the first week of my two-day week and after two days realised that I still had much work which urgently needed to be done. So I carried on for the next 98 working days – and then retired.

Before I end my memoirs I do feel I must put in a good word for the Inland Revenue Valuation Office, now the Valuation Office Agency. Often millions of pounds depend on the decisions of District Valuers and Valuation Officers, yet I have never heard of a case of corruption. I always found them a delightful lot, devoted to giving a fair, friendly and efficient service to the public. How lucky we are in this country to have such an uncorrupt, courteous and devoted Civil Service.

Today, one of my enjoyments in life is to be a member of the Seventeenth of April Club, named after the date our financial year used to end. Its membership comprises, at the time of writing, the twenty-three surviving retired equity partners of Gerald Eve (they have dropped the '& Co') and we meet for lunch at a London Club twice a year – one lunch is just for the retired partners and the other is for them and their wives or widows.

The firm is still a partnership, has sold out to no one, and seems to cope with depressions in the property market with barely a hiccup. There were seventy-four partners at the last count, spread over ten offices in the United Kingdom. I like to think that if my father, Gerald, were looking down on the firm today he would be pleased with what he saw – and with the way we had striven to carry on and live up to the timeless standards he has set. But then, as Ruskin once said:

> Quality is never an accident. It is the result of intelligent effort. There must be the will to produce a superior thing.

The End

Appendices

Appendix 1

Eric Strathon, PPRICS
From *The Times*, 31 December 1988

OBITUARY
Eric Strathon
Expert witness in valuation cases

[*Owing to* The Times *having difficulty in tracing the copyright of this obituary, I am unable to reproduce it in full, but here are brief extracts:*]

'He was one of the best expert witnesses in the country...

... during the 50s and 60s he was much in demand as one of the leading expert witnesses on valuation matters, particularly in the areas of estate duty, land compensation, business disturbance claims and planning. He gave evidence in the Lands Tribunal in many of the leading cases of the time...

In 1969 he retired from Gerald Eve & Co and was appointed a Member of the Lands Tribunal. He served in this capacity with distinction until final retirement in 1980...'

From the *Chartered Surveyor Weekly*, 26 January 1989

OBITUARY
Eric Strathon

The death has occurred of Eric Strathon, president of the Institution in 1961, a Crown Commissioner and member of the Lands Tribunal.

In a tribute, *The Times* describes him as 'one of the best valuers and expert witnesses in the country'.

He was elected to the Institution's General Council in 1951 and became a vice-president seven years later.

Eric Colwill Strathon was born on February 20, 1908. He was articled in 1925 to Mr Wilfrid Hosking of Skardon Sons & Hoskins of Plymouth and moved to London to join the firm of Thurgood Martin & Eve. He was elected a professional associate of the Chartered Surveyors Institution and an associate of the Chartered Auctioneers and Estate Agents Institute in 1929.

In 1930 he joined Charles Gerald Eve & Co, becoming a partner in 1934. In 1940 he was elected a fellow of the Chartered Surveyors Institution and was chairman of the Junior Organisation for two years.

During the early part of the Second World War Strathon was involved in the settlement of compensation following requisition of land and buildings for war purposes and war damage claims and was the author of *Compensation (Defence)* published in 1943. In the same year he joined the Royal Regiment of Artillery and served in India and at the headquarters of the 14th Army in Malaya, rising to the rank of major.

He returned to Gerald Eve after the war and, by 1962, had become a senior partner. On retirement in 1969 he was appointed to the Lands Tribunal, which he served until 1980.

He is survived by a son and daughter.

Appendix 2

J Douglas Trustram Eve, PPRICS
From the *Chartered Surveyor Weekly*, 10 April 1986

OBITUARY

John Douglas Trustram Eve FRICS (1897–1986) was president of the RICS in 1960–61. Apart from military service in both World Wars he spent his whole working life with the family firm of JR Eve & Son, which he joined as a pupil in 1915. He later became senior partner and continued to be a consultant after his retirement in 1970. Mr Eve gave many years' service to the RICS, starting with his election to the junior committee in 1923. He served on the General Council and on many standing committees, most notably that on agriculture and forestry, and was a member of the town and country planning panel. He also took a keen interest in the Institution's affairs at local level, and was chairman of the London South West branch in 1951–52. Mr Eve had a special interest in rating and was president of the Rating Surveyors Association in 1944. He was a member of council of the Town and Country Planning Association, a member of the Central Association of Agricultural Valuers, and chairman of the Farmers Club.

Appendix 3

C Gerald Eve, PPSI

From the *Journal of the Royal Institution of Chartered Surveyors*, December 1947:

OBITUARY

Charles Gerald Eve (President 1932–3) died at Hindhead, Surrey on 29th October 1946 in his seventy-fourth year. He was articled to Messrs J. R. Eve & Son, Chartered Surveyors, of Bedford in1891 and remained with them as an assistant until 1899. He then became a resident land agent in Devonshire, where he remained for ten years.

On the Passing of the Finance Act (1909–10) Act 1910, Gerald Eve entered the Valuation Office, Inland Revenue, as one of it earliest recruits and soon became a Superintending Valuer. After ten years in the Government service he joined Mr Harrison Martin (Fellow) in partnership, the firm being known as Thurgood, Martin & Eve. In 1929, the partnership having been dissolved by mutual consent, Mr Eve founded [in 1930] the firm of Gerald Eve & Co, with whom he remained until his retirement in early 1946.

Gerald Eve was a true surveyor in the very diversity of his professional appointments. Though a valuer first and foremost who devoted most of his time to rating and compensation cases, he was well versed also in the agricultural and mining branches of a surveyor's knowledge. His services as an expert witness and as an arbitrator were in wide and constant demand. There was not a single English county in which at one time or other his advice on property had not been sought. In the last fifteen years of his life he was constantly consulted by the Government when independent advice was needed on the many questions on which he could speak with professional authority.

Eve was a member of Departmental Committees on the rating of machinery in 1923, and on the Rent Restrictions Acts in 1931 and 1937. He was appointed to the Advisory Committee of the Ministry of Health on Town Planning in 1937 and in 1938 to the Lord Chancellor's Committee whose report resulted in the Landlord and Tenant (War Damage) Act of 1939. He was called as a witness by Sir Montagu Barlow's Royal Commission on the Distribution of Industry in 1939. In 1941 he became a member of the Uthwatt Committee which was formed as a result of the Barlow's Commission's recommendations regarding the State purchase of development rights in undeveloped land. The Uthwatt report will perhaps be remembered as Gerald Eve's monument, for much of it, if not actually written by him, bears the obvious marks of his genius.

On his retirement from practice in 1946 he was appointed by the Lord Chancellor, one might almost say automatically, as a member of the Tribunal formed for the purposes of the War Damage (Valuation Appeals) Act 1945, an appointment which he was compelled by failing health to relinquish a few months later.

As regards the Institution which he serves so long and so well, Gerald Eve will be remembered for his work as Chairman of the Parliamentary Committee for many years after the First World War. In that capacity he left his mark on the Rating and Valuation Act 1925, on the provisions of which many reports were made by the Institution. The same may be said of the Town and Country Planning Act of 1932 for of the sixty amendments to the Bill which had been suggested by the Institution, over forty were accepted. As a member of the Professional Practice Committee, Eve was largely the originator of the rules of professional conduct in their present form; later, during his year as President, his efforts saw them incorporated in Bye-laws and adopted by the kindred societies.

The infinite capacity of Gerald Eve for taking pains was the mark of his genius. His working day knew no limits, yet he gave a large share of his time to his work for the Institution and the profession. His help and advice were always available to his fellow members, especially younger members, whom he never failed to encourage. His contacts with others in his profession were exceptionally large. His faculty for grasping principles was never allowed to interfere with his attention to minutiae; he saw every leaf of every tree in the wood, but never missed the wood for the trees. His electric intellect could probe the recesses of highly complex legislation and illumine the obscurities with a flash of metaphor and a few simple words. So quick were the workings of his mind that mere speech seemed on occasion to be too slow a vehicle for his thoughts; in ordinary conversation he would at times search for two or three words which would do the work of ten.

Yet as a speaker on formal occasions he was outstanding. His best speeches, with their alternation of seriousness and humour, were like the play of light and shade on a fine spring day. His delivery was deliberate, his pronunciation, especially of end letters, clear and careful. Many will remember an Annual Dinner of the Institution when Royalty was present and had expressed a wish to propose a toast before rather than after the entertainers began their performance. The entertainers, for purely selfish motives, thought otherwise, and were summarily dismissed from the banqueting hall. Eve, responsible for another toast as Senior Vice-President, rose magnificently to the occasion; the entertainers were not missed, his audience was enthralled.

The family of Eve has given many distinguished members to the surveyors' profession. It has given others, as distinguished, to the Bar, to medicine, to the Services, and to scholarship. The true measure of Gerald Eve's qualities as a surveyor is that he would have achieved equal greatness in any of those fields, if he had but tried. That he chose to be a surveyor was the good fortune of his profession; now that he has gone there will be written of him in its annals the words of Virgil:

Semper honos nomenque tuum laudesque manebunt. [Your reputation, your name and your renown will always remain.]

★★★

First Ordinary General Meeting of the Session 1946–7 held at the Royal Institution of Chartered Surveyors on Monday 11th November 1946:

Before delivering his Address the President announced that, since the last Ordinary General meeting, the Institution had suffered a grievous loss by the death of Mr Charles Gerald Eve, one of the most distinguished Past Presidents of this generation.

The President, recounting Mr Eve's many attainments and his service on numerous official committees, in particular that known as Mr Justice Uthwatt's Committee on Compensation and Betterment, considered that no member of the Institution could have rendered greater service to the profession, and no man could have rendered greater assistance to the country in a professional capacity than Gerald Eve, a man who expressed his views without fear or favour.

At the invitation of the President, the meeting stood in silence as a mark of respect to the memory of Mr Eve.

Appendix 4

Professor A Stewart Eve, CBE, DSc, LL D, FRS
Reproduced from *The Times*, London, March 1948

OBITUARY
Professor A.S. Eve
Distinguished Physicist
Professor A.S. Eve, C.B.E., D.Sc, LL.D., F.R.S., who succeeded Lord Rutherford in the Macdonald Chair of Physics in McGill University, died at his home at Puttenham, near Godalming, in Surrey, on March 24 at the age of 85, as already briefly announced.

Arthur Stewart Eve was born at Silsoe, in Bedfordshire, on November 22, 1862, and attended Berkhamsted School, where he was an exhibitioner. Going up to Pembroke College, Cambridge, with a scholarship in 1881, he passed as fifteenth Wrangler in 1884, and the next year took a first class in Part II of the Natural Science Tripos. His first post on leaving the university was that of assistant master at Marlborough, where he remained for 16 years, acting for the period 1897 to 1902, when he left the college, as bursar. Then came his long connexion with McGill and with Rutherford, whose biographer Eve was to become. Appointed to the post of lecturer in mathematics in 1903, his gift of clear exposition led two years later to his promotion to an assistant professorship, and four years later again he was promoted associate professor. In 1910 the Royal Society of Canada elected him to a Fellowship.

On the outbreak of war in 1914 his eminent gift for controlling young men with firmness yet with sympathy led to his selection for a commission in the McGill Officers' Training

Corps, and he was later sent oversea as a captain in the 3rd and later the 4th Overseas Universities Companies, which acted as reinforcing units for Princess Patricia's Canadian Light Infantry, recently raised and equipped by Brigadier (then Colonel) Hamilton Gault. Valuable though his services of this kind were, his scientific knowledge was even more valuable to the war effort, and after serving as second in command of two other Canadian units he was invited in 1917 to become Director of Research at the Admiralty Experimental Station at Harwich. In the same year he was elected a Fellow of the Royal Society, and until the end of the war worked for the Admiralty, though he retained his military commission. His war services were recognized by his being made a C.B.E. in 1918, and he then returned to McGill as Director of Physics, becoming Dean of the Graduate Faculty in 1930 and retiring with the title emeritus in 1935. He was president of the Royal Society of Canada in 1929.

A busy life of teaching left him little time and energy for any permanent record of his work, though a few text-books were compiled mostly in collaboration with others. After his retirement, however, he employed his time first in writing a biography of his great colleague, Lord Rutherford, which was published two years after Rutherford's death, and secondly in writing in collaboration with C.H. Creasey, a study of the life and work of John Tyndall—a subject hitherto strangely neglected. The book appeared in 1945, and, while rightly emphasizing Tyndall's work in physics, nevertheless stresses the view, interesting as coming from a physicist, that Tyndall's most important contribution to scientific progress lay in his work on abiogenesis.

Professor Eve married in 1905 Miss Elizabeth Brooks, of Montreal, and had one son and two daughters.

Appendix 5

Dr Frank C Eve, MD, FRCP
Reproduced from *The Times*, London, December 1952

OBITUARY
Dr. Frank C. Eve
Methods of Artificial Respiration

Dr. Frank C. Eve, M.D., F.R.C.P., consulting physician to the Hull Royal Infirmary and an authority on methods of artificial respiration, died on Sunday at the age of 81.

Frank Cecil Eve was the son of Mr. J.R. Eve, and was born at Silsoe, Bedfordshire, on February 15, 1871. He was educated al Bedford School and Emmanuel College, Cambridge, where he took a first class in the Natural Science Tripos in 1893. Proceeding to St. Thomas's Hospital as a university scholar, he graduated M.B. Cambridge in 1900 and proceeded to M.D. in 1903. After holding the posts of house physician at St. Thomas's and of demonstrator of physiology at Leeds, he settled at Hull. There he became physician to the Royal Infirmary and to the Victoria Hospital for Children, and was for many years busily engaged in consulting practice. He became a member of the Royal College of Physicians of London in 1901 and was elected to the Fellowship in 1915. He took an active interest in the affairs of the British Medical Association and was formerly chairman of the East Yorkshire division of the association.

In 1932 he described in the *Lancet* a new method of artificial respiration by rocking. This method has certain advantages over the older methods of Schaefer and Silvester, and has been widely adopted in both Britain and America. Dr. Eve recently modified his rocking method and showed that it was particularly suitable for use in open boats at sea. It has since been officially adopted by the Royal Navy. Eve was a learned physician and. apart from medicine, he was a life-long student of general science and natural history.

He married, in 1911, Miss Sarah Ellice Buyers, a medical graduate of Edinburgh, who survives him together with a son of the marriage.

Appendix 6

Lord Silsoe, Bt, GBE, MC, QC
Reproduced from *The Times*, London, December 1976

OBITUARY
Lord Silsoe
Lord Silsoe, Bt, GBE, MC, QC, who died on December 3 at the age of 82, served both Church and State with equal devotion and unrivalled competence. He took on great tasks in peace and war from which lesser men would have shrunk, and carried them to fruition with cool determination and professional efficiency.

The scope of his activities was wide; he sat on important commissions—and often chaired them—of all kinds; and in his later years held such various offices as Gentleman Usher of the Purple Rod in the Order of the British Empire; president of the Kandahar Ski Club; and honorary treasurer of the Royal College of Nursing.

After a lucrative specialised profession at the Bar, he presided over a number of bodies concerned with questions relating to land, and was from 1947 to 1949 the chairman of the Central Land Board, appointed in accordance with the provisions of the Town and Country Planning Acts 1947, he had also presided over the Local Government Boundary Commission.

From his early youth he had been familiar with land in all its branches, for he was the eldest son of Sir Herbert Trustram Eve, KBE, who died in 1936, perhaps the leading authority in his day on compensation and the valuation, buying, selling and rating of real estate; and in after years father and son were not infrequently on opposite sides as expert witness and Counsel in disputes germane to those subjects. Before the death of the former, the latter had made considerable mark at the Bar, and his father used humorously to complain that he was best known as "the husband of Lady Eve (who was distinguished for her municipal and social work) and as the father of Malcolm!"

Arthur Malcolm Trustram Eve was born on April 8, 1894, and was educated at Winchester and Christ Church, Oxford. In the First World War he served with distinction with The Royal Welch Fusiliers, was at Gallipoli and was later GSO 3 to the 53rd Division and Brigade Major 159th Infantry Brigade. He was awarded a Military Cross.

He was called to the Bar by the Inner Temple in 1919, and specialized in rating and valuation and compensation cases, in which as a junior he had an outstanding practice. He took silk in 1935, and was later made a Bencher of his Inn, Reader in 1965 and Treasurer in 1966. In 1941, on the formation of the War Damage Commission, he became chairman of that body and abandoned his practice at the Bar. In September, 1944, he added to that duty that of Rehousing Chief for London. Into this work he put his accustomed energy, and by January. 1945, he was able to report that out of 390,000 cost-of-works claims received, 355,000 were then due for immediate payment. His practical interest in rehousing was shown by his giving his own home in Ormonde Gate to the Chelsea Borough Council as a home for the bombed-out.

In September, 1945, while retaining his other posts, Trustram Eve was appointed chairman of the War Works Commission under the Requisitioned Land and War Works Act, 1945, to adjudicate upon the many complicated questions regarding the future of land and buildings requisitioned by the Government during the War, and in their solution involving conflicts of interest between the State and individual owners. In November, 1947, he added yet another to his numerous posts, by accepting the chairmanship of the Central Land Board appointed in accordance with the provisions of the Town and Country Planning Acts.

He was for some years chairman of the Building Apprenticeship and Training Council. He was created a baronet for his services in 1943.

He became Third Church Estates Commissioner in 1952, and First two years later, a position he held until 1969. In this position he could put his great fund of experience and his acuteness of judgment at the service of the Church Commissioners in all branches of their work. From the start he was acutely conscious of the need to improve clerical stipends and pensions and modernize parsonage homes. He put new impulse into the Commissioners' investment policy, buying "equities" on a large scale, and so improving their resources that he was able largely by his proposals, to provide, in addition, for widows' pensions, and the building of churches in new housing areas. He uniformly backed the consolidation of the Commissioners' farm lands and the development of their urban properties was regularly pursued.

In October 1954 he agreed to become chairman of a committee set up by the Prime Minister to examine the arrangements for the administration of Crown lands and report if changes were needed. The setting up of the committee was not unexpected in the light of the Crichel Down affair. A year later Trustram Eve was appointed First Crown Estate Commissioner.

He had many other interests. At different stages in his career he rendered public service in many spheres, on the hospital and educational bodies, public appeals, Commonwealth commissions of inquiry and the Church Assembly.

With all this he was most approachable, easy in manner and in committee work, never unduly harassed without good cause, considerate and always prompt in correspondence. There was nothing of the popular conception of the lawyer's "dryness" about him, except that he displayed an occasional acidity of wit, not always appreciated by others less acute. But he suffered fools with equanimity, when he saw they were doing their best. He had little patience with those who had the brazenness to tell him how to do his job. He belonged to that select company of those who consistently put career and talents at the social and spiritual service of their fellow men. He was created a baron in 1963, taking the title of Baron Silsoe.

He married first in 1927, Marguerite, daughter of Sir Augustus Nanton of Winnipeg. She died in 1945, leaving two sons. He married secondly Margaret Elizabeth, daughter of the late Henry Wallace Robertson.

The heir to the barony is the eldest son, the Hon David Malcolm Trustram Eve, who was born in 1930.

Appendix 7

Author's talk to newcomers at Gerald Eve & Co, 1982

Notes of Talk on Philosophy of the Firm
Given to Induction Course on 12th October 1982

This is the <u>best</u> professional firm in the Country.

It was ever since my father, Gerald Eve, started it in 1930 and <u>still is today</u>.

The General Practice Division is split into agency work and professional work – <u>we</u> are the <u>best professional</u> firm of Chartered Surveyors in the Country and <u>you</u> are going to help us keep it so.

What is professionalism? Let us trace the history…Gerald Eve started the firm in 1930 at the age of 53 [57]. He had made his name as an expert witness in rating and compensation and other appeals before the Courts. He had a good brain, and integrity – and whereas other expert witnesses were renowned for their persuasiveness or dogmatism, he made his name for omniscience – literally all knowing. In other words, he did his homework, thus the firm got a reputation for brains, clear thinking and doing a sound thorough professional job.

Professionalism means first <u>integrity</u>. This includes intellectual honesty. If a client offers you a £10,000 fee for an asset valuation but wants first to ensure you will value his assets at £10 million, you say "No, it may only be £8m", and if he goes somewhere else and a year later you see in the papers that firm has valued his assets at £10 million, you will grin and bear it. Your first duty is to be true to yourself, your conscience and your profession. Similarly, when you are negotiating with the

Valuation Officer, remember areas are <u>facts</u> and your areas should be just as likely to be <u>more</u> than the Valuation Officer's than less, and in compensation as likely to be less than the Valuation Officer's as more. These are <u>facts</u>. The game does not start until one gets to <u>values</u> and <u>opinions</u>.

Next after integrity "put the client first". No need to be hypocritical about this; you do actually make <u>more</u> money if you do this, and if you are going to be in business for more than just a day or so. Forget the fee, that is the <u>partner's</u> job, either before they start or after they have finished. Do the job well – superbly. If you need to go to Ashby-de-la-Zouch three times to get an assessment correct, do so. Never mind if we lose on that job. We can afford that. There is only one thing we <u>cannot</u> afford – a <u>botch</u> job. Our reputation is the most precious thing we have.

Negligence

It is one of the jokes in the profession, and every surveyor knows it, a client never wants a <u>structural survey</u> just "a quick look". No client wants a valuation just "a rough idea of what it's worth". Woe betide you if this leads you to do other than a first class thorough job. If you don't you may be sued for negligence, even if you did the job for nothing. Remember, we are not a limited liability company and if we leave a nought off and value Marks & Spencers at £100 million rather than a billion, we will be sued for the £900 million, and if the partners cannot pay (it is rather a lot), they can sue you, for you too have a <u>duty of care</u> to the partners. This may not disconcert those of you who are penniless, but those who are the proud possessor of a bicycle might mourn its loss.

If one makes a mistake it is never a defence to say "I was in a hurry". Remember, <u>all</u> important jobs are done in a hurry, often late at night. This is when you are most likely to make a mistake and when you <u>must</u> check work. Get the noughts right, e.g. 19×21 must be about 400 not 4,000.

So what is not tolerated in this firm is <u>sloppy maths</u>. Nor is <u>sloppy thinking</u> or <u>sloppy language</u>. If you have a problem, define it, get the facts, consider the various alternative solutions – and choose the best one.

We don't want any "on-going jargon situations at this present moment in time". Sloppy language permits sloppy thinking. Remember, the best brains use the simplest words, e.g. Lord Denning. When you are writing letters, or dealing with clients remember we are not in the <u>property</u> business, we are not in the <u>valuation</u> business, we are not in the <u>planning</u> business – we are in the <u>people</u> business. It is no good doing a marvellous job technically, if you fail to keep the client informed, fail to explain matters to him and fail to deal with his problem from <u>his</u> point of view. So when you write a letter, write naturally, as if you were talking. Don't use commercialese and don't write "the site is approximately 1.973 hectares" – say "it is about 5 acres". When you initial a letter for signature by a partner or associate, read it carefully including date, address, heading and check the maths and figures. Make sure your initials <u>mean</u> reliability – like the St. Michael label. "Any reason why I should read a letter more slowly than you. Is your time more valuable?"

Finally, I would refer to the quotation I have handed out to you "Quality is Never an Accident. It is the result of intelligent effort." You all have the intelligence or you wouldn't be here – you all have the requisite 'O' and 'A' levels. It is up to you to make the "effort" and to have the "will to produce a superior thing."

That is the way, the only way, we can maintain the standards and reputation of this firm. I am sure we can count on you all to do your best to maintain and indeed improve these standards and I hope you will enjoy yourself here trying to achieve this. Take a pride in your work, set yourself the highest standard, don't be content with second best.

I will end as I have started. Remember, we are the best professional firm of Chartered Surveyors in the Country and you are part of it. Some of you may retire from this firm as a partner, others may leave for pastures new as soon as they get their TPC. Either way, wherever you go in the world, remember never, never, whatever the pressure or temptation, never lower your standards.

If you have any questions I think it is best that you bring these up over drinks at lunch as we are now going to adjourn.

Thank you very much.

H.M. Eve

Appendix 8

Draft letter to Lord Denning, 1982

Dear Lord Denning,

I have over the years enjoyed reading your decisions and, recently on my retirement, your books. Your latest one, *Landmarks of the Law*, prompts me to write to you about a very sad decision of the House of Lords in the case of *K Shoes v Hardy (VO) and Westminster City Council;* sad because the decision resulted in all London's Regent Street traders being assessed unfairly high; sad because it will, between now and the next rating revaluation in England and Wales (which is probably at least five years away) cause great injustice to ratepayers appealing against their new or altered assessments.

Briefly, all Valuation Officers when making their 1973 Valuation Lists were instructed to send out their Return Forms between June 1969 and June 1970, all valuation work to be done between April 1969 and December 1971. Later rents were generally not had regard to because the VO would not have time to reflect them in the Valuation List as a whole. By December 1971 provisional totals for each class had to be given to the Department of the Environment.

As a result, all Valuation Lists in England and Wales have in fact a prevailing level of assessments equal to 1971 rentals or earlier. Every VO knows this but will not tell you, and every rating surveyor in private practice knows this and will tell you so.

However, in the City of Westminster, the VO valued the residential properties first, then valued Bond Street and Oxford Street in 1971 at values current then, and finally Regent Street in April to August 1972 at rentals current during that period.

I did some research as to the tone of the List in Westminster and found it coincided with early 1970 values. This evidence as to the tone date of the List was derided in the Court of Appeal and House of Lords, but (a) it was the <u>only</u> evidence given on tone date; (b) the evidence of the relationship of rents and assessments was taken from 182 rents and assessments in 55 different streets throughout the Valuation Area; (c) what is better evidence than an agreed schedule of 182 related rents and assessments? Surely one with 183 or more, rather than no schedule at all and no evidence of rebuttal.

Because of this research, the VO agreed to reduce over half the assessments in Bond Street and three-quarters of the assessments in Oxford Street. Finally, when it came to Regent Street and I proved that the level of assessment was thirty per cent above tone, because the street had been valued last of all using the latest rents, the Revenue, under pressure from the Westminster City Council, retired the VO who had agreed the Oxford Street and Bond Street reductions. They put in a new VO who could, and did, say at the Lands Tribunal that he had no knowledge of the level of assessment in Westminster or of the reasons for the Oxford Street and Bond Street reductions.

The Treasury Solicitor and the Court then took the line that there was no such thing as tone of the list and that if there were it would be impossible to find it. They also contended that the level of assessment in the Valuation list should be 1 April 1973.

I took this case myself in the Local Valuation Court and gave evidence in the Lands Tribunal. In the former I obtained a limited but insufficient reduction in the basis of assessment and in the Lands Tribunal managed to no more than hold onto this reduction.

In the Court of Appeal and the House of Lords our appeals were dismissed. Frankly, I did not approve of the way our case was conducted and said so – which made me unpopular.

I had hoped to use my cousin, Lord Silsoe (one of the authors of *Ryde on Rating*), in this case, but he was otherwise engaged up at The Maltings, and to have the case tried by you in the Court of Appeal. I felt sure that there you would again apply the principles you adopted in the *Peachey* case – but this time supported by an Act which, so David Silsoe and I thought, provided for statutory tone of the list – rather than statutory valuation of Regent Street at 1 April 1973 rental values.

If the law says the VO can accurately predict values up to three years ahead, the law is an ass. The law thinks just that in the *K Shoes* case and I feel sure that as a lifelong upholder of the law you would not wish it to continue to be an ass – causing great injustice to hundreds of ratepayers; for the decision in this case can cause great injustice, because there was an unprecedented property boom between 1970/1 and 1973 resulting often in a fifty per cent or more increase in shop rental values.

If you would like more facts and figures I can arrange for you to have them. They are all at my office – but I hesitate to submerge you with paper.

What can you do about all this, you may say?

Well, first, since I have always respected your judgment and your creative approach in dispensing justice, it would please me no end to hear you say that you agree the decision is wrong.

Second, you could introduce a bill in the House of Lords to put matters right – or use your influence to get someone else to do so.

Third, you could use your influence to ask the Government to introduce a bill to legislate for statutory tone of the list. If one refers to Hansard, it seems that that is what the Government thought it was doing when debating the bill which became the 1971 Act – and indeed what most of us rating surveyors, and some distinguished rating lawyers, thought had in fact been done. Unfortunately, the expression 'tone of the list' never got further than a marginal heading – which, as you know, has been held not to be part of the Act.

I would be delighted to have a chat with you about it if you could spare the time – either at Whitchurch or in the House. I know that there is no one busier than you, but I am a great believer in the saying 'if you want something done, ask a busy man'.

Kind regards.

I have met you and your wife once or twice before – once, I think, at Hicks Beach's sesquicentennial cocktail party.

I hope you are keeping well.

Yours very sincerely,

Hilary M. Eve

Index